Spaceplanes

Matthew A. Bentley

Spaceplanes

From Airport to Spaceport

Matthew A. Bentley
Rock River
WY, USA

ISBN: 978-0-387-76509-9 e-ISBN: 978-0-387-76510-5
DOI: 10.1007/978-0-387-76510-5

Library of Congress Control Number: 2008939140

Printed on acid-free paper

springer.com

This book is dedicated to the international crews of the world's first two operational spaceplanes, Columbia *and* Challenger, *who bravely gave their lives in the quest for new knowledge.*

Challenger

Francis R. Scobee
Michael J. Smith
Ellison S. Onizuka
Ronald E. McNair
Judith A. Resnik
S. Christa McAuliffe
Gregory B. Jarvis

Columbia

Richard D. Husband
William C. McCool
Michael P. Anderson
Ilan Ramon
Kalpana Chawla
David M. Brown
Laurel Clark

Contents

Preface

Spaceplanes are an important part of our future in space. It is the aim of this book to show how spaceplanes of the near future will be developed, how they compare to conventional aircraft and launch vehicles, and how advanced spaceplanes will eventually become the dominant form of space access. In this book, you will find material on airplanes and rockets, in addition to rocketplanes and spaceplanes. The idea here is to allow true comparisons to be made, because only by comparing these modes of transport can one develop a good understanding of the subject.

The idea of winging one's way through airless space might seem strange. But a thoughtful analysis of every stage of flight, including the return trip, reveals that spaceplanes have a future not just in near-Earth space but also farther afield.

The idea of lunar and interplanetary spaceflight using spaceplanes begins to make more sense when several important factors are recognized. These factors include (1) the spaceplane's large propellant tanks, which would make it the ideal space tanker, (2) inflatable space habitats that could easily be carried on and deployed from spaceplanes, (3) the low *G* entries spaceplanes can make into planetary atmospheres, and (4) the fact that they can fly efficiently through any atmosphere, including the atmospheres on Mars, Venus, and Titan. Of course, the main factor is the reusability built into every spaceplane, which will make spaceplanes the space workhorses of the future. But let us start at the beginning, and get a handle on a few basic definitions.

What is involved in spaceflight? To understand this complex question, it behooves the wise reader to understand what precisely is meant by *space* and what precisely is meant by *flight*. Entering space does not necessarily mean entering orbit. In fact, the energy requirements for entering into low-Earth orbit are some 30 times those required for reaching space height, defined as 100 km above sea level. It is possible to orbit Earth at an altitude of 160 km, yet a stationary object – relative to Earth – placed 1.6 million km away will still fall to the ground as assuredly as any dropped object. The only exception is that it may fall onto the Moon first, if that celestial body happens to get in the way. The reason the X-15 and SpaceShipOne could not stay in space was not because they had not gone far enough, but because they had not gone fast enough. But back to the subject at hand, space itself.

The word *space* conveys the idea of room. *The American Heritage Dictionary of the English Language* defines space as room, and tells us the Indo-European root, *rewe*, also means "to open." Space certainly opens up new possibilities, broadens the mind, and expands the sphere of human activity. It gives us room to grow. It promises new technologies with improved capabilities. It unlocks the door to the future.

The idea of *flight* has three separate meanings: (1) to take wing above Earth, (2) to flow or run away rapidly, and (3) to float. These meanings are exemplified in the following kindred words: *fly*, *flew*, *flown*, *fowl*, *flutter*, *flee*, *fleet*, *flow*, *flood*, and *float*. The last term involves buoyancy in a fluid medium, such as water or air. Boats and balloons each rely on the same principle to stay *afloat*. And of course, spaceflight involves *floating* in space. Each of these terms can trace their origins back to the same Indo-European roots.[1] The word for river, a flowing stream of water, also stems from the same root in several Germanic languages – the Swedish *flod* and German *Fluss* are two examples. And water is a compound of hydrogen and oxygen, the best rocket propellants known. Based on this simple etymology, the term *spaceflight* conveys the images of *taking wing* above Earth, an *airflow* over a lifting body, a *rapid run* into a *roomy realm*, a *fleeing* from the bonds of gravity, a *flood* of propellants into and a *flow* of high-speed exhaust from an engine, and finally, *floating* in space. Spaceplanes seem, etymologically at least, particularly suited for spaceflight.

There are a few other terms you should be familiar with before we dive into the meatier sections of the book. There is also a glossary of terms in the Appendix.

An *aircraft* is any craft built to operate in air, including balloons, blimps, gliders, airplanes, helicopters, and rocketplanes.

An *airplane* is a fixed-wing aircraft that flies by aerodynamic principles.

A *ballistic missile* is typically a rocket-powered aerodynamically shaped cylinder used to lob a projectile. It may have any range, and it may or may not enter space.

A *rocket* is a ballistic missile intended for spaceflight. It can also refer to a rocket engine, as in "rocket-powered launch vehicle."

A *rocket engine* is the device that produces the thrust for the rocket, launch vehicle, or spacecraft. Solid propellant rockets are called "rocket motors."

A *rocketplane* is an airplane powered by a rocket engine. It may or may not have the capability of entering space.

A *rocketship* is a spaceship powered by a rocket. The term is often used to describe spaceships that do not require separate launch vehicles. They are nonmodular, completely self-contained, and sport huge fins that serve to balance them when they land on other planets.

A *spacecraft* is any craft built to operate in space, including artificial satellites, planetary probes, space stations, spaceships, and spaceplanes. Spacecrafts have historically relied on launch vehicles to enter space.

A *space launch vehicle* is the rocket used to launch a spacecraft. It usually consists of two or more stages that drop off when their propellants are expended.

A *spaceplane* is an airplane that flies in space, or a fixed-wing spacecraft.

A *spaceship* is any manned space vessel, especially one intended for shipping crews or cargo.

A *space shuttle* is a spaceship that shuttles back and forth, typically between a planet and a space station, or between a planet and a moon. It may or may not be winged.

A *space station* is a large manned spacecraft, usually stationed near a planet.

Claims have been made that the Space Shuttle fleet operated by NASA are not true spaceplanes. In fact, the Space Shuttles are not fully developed spaceplanes, but they are true spaceplanes nonetheless. Likewise, both the X-15 and SpaceShipOne are true spaceplanes. They represent the first baby steps on the road to truly usable spaceplanes.

This book is about winged space vehicles. However, the reader should not take the term "winged" too literally, because the ultimate shape of spaceplanes has not yet been determined. It may very well turn out to be a "wingless" lifting body, or a blended wing-body of some kind. Also, the term "fuel" may sometimes refer to both the actual rocket fuel and the oxidizer necessary for operating rockets in space. The context should make it clear that sometimes "fuel" means "propellant," but we have used poetic license at times to make for a more enjoyable reading experience.

Rock River, WY Matthew A. Bentley

Reference

The American Heritage Dictionary of the English Language. Dell, New York, 1976.

Acknowledgements

Heartfelt thanks to
David Ashford of Bristol Spaceplanes, for graciously allowing me to quote from his book *Spaceflight Revolution* and use illustrations of his designs, as well as for sharing certain technical information.

Alan Bond of Reaction Engines Ltd., for providing excellent images and technical information on the Skylon spaceplane and its advanced propulsion system.

Fletcher Manley, my uncle, for taking the time and providing the expertise to convert each of the book's images from JPEG to TIFF format for the publisher.

Dr. Pamela Menges of Aerospace Research Systems, Inc., for providing specifications and images of the Ramstar orbital spaceplane.

The National Aeronautics and Space Administration, for providing the vast majority of excellent photographs in this book, and for providing authoritative historical and technical information on the many projects it and its predecessor, NACA, has worked on over the years.

Vassili Petrovitch, for allowing me to use illustrations of the Buran, BOR, and Spiral spaceplanes from his website.

Vidhya Rajakoti, for overseeing the final production of the book in a precise, professional, and punctual manner.

Maury Solomon and Turpana Molina, my editors at Springer, for keeping me on track, prodding me on, and providing crucial editorial aid.

Dr. Raymond Wright of Spacefleet Ltd., for his friendship, and for supplying illustrations and information on the SF-01 spaceplane.

Note to Reader

This book is being marketed to potential readers in the Americas, Europe, and around the world. Because of the differing engineering units used in the aerospace industry on either side of the Atlantic, both sets of numbers will be used throughout the book, choosing whichever system seems to suit the particular example. Thus, discussions of the Apollo program will use US customary units, while those of the Soyuz or Ariane programs will use strictly metric units. For the benefit of all readers, it is good to be familiar with a few basic values common in the aerospace field. These are given in the following table, in both their US customary and international metric system equivalents:

Light plane altitude	10,000 ft	3 km
Airliner altitude	50,000 ft	15 km
Space altitude	330,000 ft (62 miles)	100 km
Orbital velocity	17,500 mph (5 miles/s)	8 km/s
Escape velocity	25,000 mph (7 miles/s)	11 km/s
Low earth orbit	100–500 miles	200–800 km
Earth to Moon	238,000 miles	383,000 km
Sun to Earth	93,000,000 miles	150,000,000 km
Sun to Mars	142,000,000 miles	229,000,000 km
Earth to Mars	370,000,000 miles[a]	600,000,000 kma
Galaxy diameter	100,000 light-years	30,000 parsecs

[a] Assuming Hohmann transfer orbit, which takes the spacecraft halfway around the Sun.

As a rule of thumb, it is a good idea to memorize the "easy" numbers. An educated space-savvy resident of Planet Earth should know that space officially begins at 100 km, that low Earth orbit is centered at about 200 miles – or 300 km – above sea level, and that the diameter of the Galaxy is 100,000 light-years (74 trillion Earth diameters!). It is also useful to know that the speed of light is 300,000 km/s, and that a light-year is the distance that light travels in one tropical Earth year. This amounts to just under six trillion miles. It is also vital to know that the difference between escape and orbital velocities is a much smaller value than that of orbital velocity itself, and that planetary atmosphere entry speed is roughly the same as the escape velocity, if the spacecraft is entering from interplanetary or cislunar space.

When numerical values are spelled out, as in the cases just cited, standard American usage will be employed, so that progressing by factors of 1,000, the sequence is thousand, million, billion, trillion, quadrillion, quintillion, etc.

Since the kilogram is technically a unit of mass, while the pound is a unit of weight, this book will typically use these terms in this way, to avoid ambiguity. The corresponding units of weight and mass in the respective systems are newton and slug, which are rarely used in common parlance. For those who seek further information, please see the mathematical appendix.

Chapter 1
Rocketplanes at the Airport

Rocketplanes have been around far longer than many of us groundlings may realize. German rocket gliders were flying in the 1920s, and winged space-bombers have been on the drawing boards since the 1930s. Both rocketplanes and winged rockets were flown during World War II, and rocket-research aircraft flew from the late 1940s until the mid-1970s. Rocket-powered "lifting bodies" flew in the 1960s and 1970s, gathering much needed aeronautical data. All of these efforts led directly to the US Space Shuttle, which since 1981 has flown over 120 missions into low Earth orbit, each time landing on a runway as a real spaceplane. The latest working example of a real spaceplane is SpaceShipOne, which flew several suborbital missions in 2004 above the 100-km "magic altitude" of space. The idea of using a winged vehicle to gain access to space shows no signs of stalling out. The vehicles themselves have had their share of mishaps, to be sure, but the idea flies on.

The history of the development of the spaceplane is one that stretches back to the early years of the twentieth century, and is intimately tied to the history of the airplane itself. Visionaries had foreseen an inevitable evolution of the airplane into a vehicle capable of spaceflight since at least the 1930s. Rocket engineers, by contrast, were developing wingless missiles in their endeavor to reach space by purely ballistic means. We will address the fundamental reasons for this later on. Yet, those in favor of the spaceplane concept recognized, from the outset, that a rocket engine of some kind could be "married" to a winged airframe to create a long-range vehicle capable of space operations. This concept took several forms, reflecting the initial design approach. One avenue of attack was to incorporate a rocket engine into a highly streamlined airplane, as in the X-series of research aircraft, and see how high and fast you could go. Another approach was to vertically launch a ballistic vehicle that could glide back to Earth, a method first tried with the V-2 rocket and routinely used by the US Space Shuttle. Both approaches have yielded varying degrees of success, and the debate about which approach is best continues to this day. But to properly tell the saga of the spaceplane and its progenitor, the rocketplane, we have to start with the first powered airplane.

M.A. Bentley, *Spaceplanes: From Airport to Spaceport*,
doi:10.1007/978-0-387-76510-5_1, © Springer Science+Business Media, LLC 2009

The Wright Flyer

The first powered, piloted, controlled, heavier-than-air flight took place at about 10:35 on the morning of December 17, 1903, at Kitty Hawk, North Carolina. With the help of a 60-foot guide rail, this first flying machine lurched into the air, flew erratically for a distance of 120 ft, and came to Earth after just 12 s. The *Wright Flyer I* had averaged 10 ft/s, or 6.8 mph groundspeed. By today's standards, this sterling accomplishment of Wilbur and Orville Wright seems almost laughably primitive. Yet they accomplished something of lasting significance in the annals of aerospace history. Before this first powered flight, they had made some 1,000 glide flights, and were therefore experienced pilots. At the time, there had been no engine to power their *Flyer*, and so they simply built their own, an accomplishment in itself. This is an important fact, for improvements in the field of aviation are almost always accompanied by improvements in power. The Wright brothers flew the *Flyer I* three more times that first morning, the last time covering 852 ft in 59 s (Fig. 1.1).[1]

Fig. 1.1 Orville Wright makes the first powered, controlled, heavier-than-air flight in human history, December 17, 1903, Kitty Hawk, North Carolina (courtesy NASA)

Fig. 1.2 Fully suited aeronaut prepares to take Wright Apache to high altitude in 1928. The 18-cylinder radial engine was a significant advancement over the *Wright Flyer* (courtesy NASA)

There are certain discontinuities in engineering, analogous to the kind found by geologists in layers of ancient rock. These discontinuities represent a gap in the fossil record, or a jump in technology. Replacing the horse by the engine was one such discontinuity. Making the switch from airships to airplanes was another. Yet a third was the jump from the reciprocating engine to the turbojet. When the Wright brothers mounted chain-driven propellers and a 12-hp engine to the frame of *Flyer I*, they had just made one of these leaps of progress. And there was no going back.

The process of refining and improving the design of the first airplanes started almost immediately after the first flights. It was not long before the Wright brothers could fly around the patch with confidence. They controlled their craft through a unique wing-warping scheme, rather than by the use of ailerons and elevators. Later, the invention of the radial engine (Fig. 1.2) and the development of the multiengine airplane would extend the power and range of airplanes. World War I saw the first use of the airplane in combat, and the first aerial dogfights.

Rocket Men

As the airplane continued to see slow but continuous improvements throughout the early years of the twentieth century, others were thinking beyond the atmosphere. They were considering the future of spaceflight by rocket. The main characters in this story were the reclusive Russian schoolteacher Konstantin Eduardovich Tsiolkovskiy, the German genius Hermann Oberth, and the American physicist Robert Hutchings Goddard. These three men, early on, represented the three nations that would play pivotal roles in the fascinating saga of the spaceship over the course of the ensuing century (Figs. 1.3 and 1.4).

It was Tsiolkovskiy who first wrote about the value of liquid propellants as the modus operandi of future spaceships. By 1895, he recognized that, unlike solid-propellant rockets, liquid-propellant rocket engines could be controlled, throttled, turned off, and turned back on. This would enable much greater flexibility in the execution of flights in the cosmic realm. Liquid fuels such as hydrogen, with their low molecular weights, would give the required impulse to

Fig. 1.3 Rocket pioneer Hermann Oberth (1894–1989) lived to see men walk on the Moon and spaceplanes land on Earth (courtesy NASA)

Fig. 1.4 Dr. Robert H. Goddard (1882–1945) at work in 1924 at Clark College, Worcester, Massachusetts, 5 years after writing his Smithsonian-published paper "A Method of Reaching Extreme Altitudes" (courtesy NASA)

these spaceships. Oberth's contribution was the 1929 publication of a detailed treatise entitled *Wege zur Raumschiffahrt* (*Ways to Space Travel*). This was a greatly expanded version of his 1923 work *Die Rakete zu den Planetenräumen* (*The Rocket to Interplanetary Space*). But it was the American, Robert Goddard, who first turned the physics of the liquid-propelled rocket into a practical flying device. On March 16, 1926, from a field at his Aunt Effie's farm near Auburn, Massachusetts, he launched the first liquid fuel rocket. The device did not look much like a rocket (Fig. 1.5), but it demonstrated the rocket principle amply enough, rising to an altitude of 41 ft and crashing to the ground 2½ s later. It reached a speed of 64 mph.

Silbervogel: The Silverbird

By the 1930s visions of spaceplanes were dancing in the heads of certain forward-thinking dreamers. The *Silbervogel* concept was one such vision, held by the Austrian engineer Eugen Sänger. Rather than starting with vertically launched rockets to be ballistically *lobbed* into space, as the conventional rocketmen of his day were contemplating, Sänger was interested in designing a real *Raketenflugzeug* – a rocket-airplane – for *flying* into space. He may have been the first to appreciate

Fig. 1.5 Robert Goddard with the first successful liquid-fueled rocket, March 16, 1926. The device used gasoline and liquid oxygen propellants (courtesy NASA)

that a pilot's spaceship must be flown, not thrown, to its destination. In this respect, he was well ahead of his time. In 1933 he published *Raketenflugtechnik*, often literally mistranslated as *Rocket Flight Engineering*, but more accurately translated as *Rocketplane Engineering*.

The Silverbird concept required more weight than simple rocket designs, because of the need for a structurally sound wing root and landing gear. Sänger immediately recognized the weight penalty this imposed, and he set about solving the problems it presented. Launch would be from an auxiliary track rather than from an ordinary airfield. This would allow a separate rocket sled to provide the initial momentum, thereby relieving the vehicle of extra weight. Thereafter, the pilot would fly *Silbervogel* up through the atmosphere, gaining maximum speed at the edge of space. The idea was then to fly a series of roller-coaster-like "skips" entering and exiting the upper atmosphere on the way to the ultimate destination. The wings would thus provide a much longer range, even if it meant a higher take-off weight than in the case of a simple ballistic design. It was believed that the craft could reach a point half-way around the world from the launch site and release a load of bombs anywhere in between – hence, the name "antipodal bomber."

Sänger and his future wife, the mathematician Irene Bredt, collaborated on the *Silbervogel* design during the latter years of World War II, although no flight hardware was ever manufactured. This may have been an unfortunate circumstance for the development of spaceplanes in general, but certainly fortunate for New York City, which would likely have been its first target.

German Rocket Gliders

The first rocket-powered airplane was flown on June 11, 1928, at the *Wasserkuppe*, a popular gliding spot in the Rhön Mountains of Germany. Using two black-powder solid rockets fired in sequence, Fritz Stamer became the first rocketplane pilot in history. The rocket glider *Ente* was catapulted into the air from the 950-m mountain, and started its rocket motors with a pilot-activated electric switch. Two flights were made. The first time, the rockets were ignited in sequence, propelling the craft at least 1.2 km. On the second flight it was decided to start the rockets simultaneously, but this caused an explosion and set the wings on fire. From a height of 20 m, test pilot Stamer still managed to bring the craft to the ground and escape unharmed, but the rocketplane itself was a total loss.

Little more than a year later, on September 30, 1929, Fritz von Opel flew a specially built rocket glider, the RAK.1, before a crowd of onlookers near Frankfurt-am-Main. The small craft was powered by a cluster of 16 solid rockets. He flew 1.5 km and made a hard landing that damaged the glider, but the flight again proved that rockets could propel a fixed-wing aircraft.

The first liquid-propellant rocket-powered airplane flew on July 20, 1939, at the Peenemünde rocket range. It was the Heinkel He 176, piloted by Erich Warsitz, and used a Walter diluted hydrogen peroxide (H_2O_2) and hydrazine-fueled engine. Prior to this, there had been numerous experiments with solid-propellant rocket-powered vehicles, both on land and in the air. But these could not be throttled or controlled. The He 176 was made of wood and weighed 900 kg empty, 1,620 kg loaded. It was 5.2 m long, spanned 5 m, and stood 1.5 m in height. It had a powered endurance of only 50 s.[2–4]

As is often the case, wartime accelerates the development of new technology, and World War II was no exception. For Germany, this took the form of the Messerschmitt Me 163 *Komet*, the first *operational* rocketplane in human history (Fig. 1.6). Like its immediate predecessor the He 176, it used liquid propellants in a Walter rocket engine. The Me 163 set a world airspeed record in 1941 of 1004.5 km/h. The *Komet* played the role filled in later years by the surface-to-air missile. It would take off with the help of a wheeled dolly that remained on the ground, make a steep ascent to high altitude, and level off in preparation for a high-speed attack run on American or British bomber formations. Within 8 min, all propellants would be gone, turning the *Komet* into a glider. At this point, the pilot had to set up the correct glide path, find his home airfield, and land by means of a single tail wheel and extendable skid. As a combat aircraft, the Me 163 had the dubious distinction of killing more of its own pilots than the enemy, typically during landing. But as a rocketplane it earned its place in history. It was the first rocket-powered airplane to

Fig. 1.6 The Luftwaffe's Messerschmitt Me 163 *Komet*, the world's first operational rocketplane (courtesy United States Air Force)

be used operationally, officially entering service for the Luftwaffe in 1944. Although other flying test beds had been fitted with rocket motors prior to the Me 163, this was the first time a rocket-powered aircraft had been designed for regular operational use.

American Rocketplanes

With new airframe designs being developed for the more powerful jet engines, the years following World War II became a Golden Age for the rocketplane. The epicenter of rocketplane research was Edwards Air Force Base, California. Unlike the strictly subsonic German rocket gliders, the new American vehicles were built for supersonic speeds and high-altitude research. Their goal was to test the limits of aeronautics at those speeds and altitudes, and to test the handling characteristics not only at those limits but throughout the entire flight envelope.

The Bell X-1 (1946–1958)

It took a rocketplane to break the sound barrier. The Bell XS-1 (Fig. 1.7), piloted by US Air Force Captain Chuck Yeager, accomplished this feat on October 14, 1947. His rocketplane, nicknamed *Glamorous Glennis* after his wife, was modeled

Fig. 1.7 The Bell X-1 in flight, 1947. Notice the shock diamonds in the plume caused by the supersonic exhaust (courtesy NASA)

Fig. 1.8 Chuck Yeager with his *Glamorous Glennis*, the X-1 rocketplane that "broke the sound barrier" on October 14, 1947 (courtesy USAF)

on a .50-caliber machine gun bullet, with stubby wings attached (Fig. 1.8). The X-1 program was a joint effort between the National Advisory Committee for Aeronautics (NACA) and the United States Air Force (USAF). The Experimental Sonic XS-1 burned ethyl alcohol and liquid oxygen (LOX) in its four-chamber 6,000-lb thrust (sea level static) XLR-11 rocket engine, but it could not take off by itself, which would have wasted precious propellant. Instead, a modified B-29 bomber (Fig. 1.9) carried the little rocketplane under its wing and released it at 21,000 ft, where it ignited its engines and flew into history. Other airplanes had flown faster than sound during high-speed dives, but the X-1 was the first to do so in level flight. Progressing through five different versions culminating in the X-1E, this rocket-powered pioneer made a total of 238 test flights from the mid-1940s until the late 1950s.[6,7]

Fig. 1.9 The X-1 "baby" rocketplane with its B-29 mothership, "Fertile Myrtle." See the stork carrying the baby on Myrtle's nose (courtesy NASA)

The Douglas Skyrocket (1948–1956)

The significance of the Douglas D-558-2 was that it was the first airplane to reach twice the speed of sound. The Skyrocket (Fig. 1.10) was sponsored by the NACA and the US Navy, and there was keen competition with the NACA/USAF team and their X-1 program to be the first with this milestone. To accomplish it, pilots and engineers spared no effort. By adding nozzle extensions, waxing the wings, and chilling the alcohol fuel to increase its density, it was hoped that the Skyrocket could be the first to reach Mach 2. NACA test pilot Scott Crossfield reached this goal on November 20, 1953, just a little over 6 years after Chuck Yeager pierced Mach 1 in the X-1. This milestone was accomplished at an altitude of 72,000 ft.

The Douglas Skyrocket used the same engine as that of the X-1, with its own Navy designation (XLR-8), but it had a 35° swept wing rather than the straight wing of the X-1. Its wing area was 175 sq ft, compared to 130 sq ft for the X-1. Three Skyrockets were built for the USN, and together they made 313 flights during the 8-year test program. The Skyrocket was both jet- and rocket-powered. The first Skyrocket was initially powered only by a 3,000-lb thrust turbojet and carried 260 gallons of aviation gasoline. This was later replaced by the 6,000-lb thrust rocket engine. The second Skyrocket had the rocket engine only. And the third vehicle used both jet and rocket engines at the same time. Rocket propellants were LOX and diluted ethyl alcohol. The third Skyrocket, which flew 87 times, was able to carry 260 gallons of aviation gasoline, 179 gallons of LOX, and 192 gallons of rocket fuel.[8]

Fig. 1.10 The Douglas D-558-2 *Skyrocket* is dropped from P2B-1 mothership (Navy version of the USAF's B-29 *Superfortress*) in 1956. Although technically not an X-plane, it nevertheless contributed much to aeronautics research (courtesy NASA)

The Bell X-2 Starbuster (1952–1956)

Following the transonic and supersonic research mission of the X-1 up to about Mach 2, the X-2's goal was to investigate the region above Mach 3. The X-2 program was a joint effort between Bell Aircraft, the US Air Force, and the NACA. It was plagued with problems, and ultimately took the lives of three men. Bell built two swept-wing aircraft, each containing a Curtiss–Wright XLR-25 throttleable liquid-fueled rocket engine with two thrust chambers. One was a 5,000-lb chamber and the other, a 10,000-lb thrust chamber. The X-2 could therefore develop anywhere from 2,500 to 15,000 lb of thrust. It was fitted with an escape capsule in the form of a nose that could be jettisoned with a stabilizing parachute. The pilot would then bail out when he reached a safe altitude. A Boeing B-50A mothership would release the X-2 at around 30,000 ft, as seen in Fig. 1.11.

During a captive-carry flight to check the LOX system over Lake Ontario in 1953, there was an in-flight explosion resulting in the loss of X-2 pilot Jean L. "Skip" Ziegler, B-50 crewman Frank Wolko, and the X-2 itself, which fell into Lake Ontario and was never recovered. These were the first casualties of the problem-ridden program. But the Air Force still had one X-2 left.

USAF Captain Iven C. Kincheloe earned the title "first of the spacemen" when he flew the X-2 to 126,200 ft on September 7, 1956, a new record. Only weeks before this flight, NACA had requested use of the X-2, but the Air Force wanted to make one more flight in an attempt to reach Mach 3.

On September 27, 1956, Air Force Captain Milburn G. "Mel" Apt flew the rocketplane to a record 2,094 mph, or Mach 3.2, at 65,500 ft. Although warned not to make any abrupt control movements above Mach 2.7, he turned sharply back

Fig. 1.11 Bell X-2 ignites rocket engine after drop from B-50 mothership, 1955 (courtesy NASA)

Table 1.1 Early American rocketplanes

Rocketplane	Length	Wingspan	Empty wt. (lb)	Flights	Max. speed (mph)	Max. Mach	Max. altitude (ft)
Bell XS-1	30 ft 11 in.	29 ft	6,784	157	957	Mach 1.4	71,902[10,11]
Bell X-1A	35 ft 7 in.	28 ft	6,880	26	1,650	Mach 2.4	90,440[12,13]
Bell X-1E	31 ft 10 in.	22 ft 10 in.	6,850	26	1,471	Mach 2.2	73,050[14,15]
D-558-2	42 ft	25 ft	9,421	313	1,250	Mach 2.0	83,235[16]
Bell X-2	37 ft 10 in.	32 ft 4 in.	12,375	20	2,094	Mach 3.2	126,200[17,18]

towards Edwards while still above Mach 3 and encountered a dangerous control divergence known as inertial coupling. The vehicle tumbled out of control, and both he and the only remaining X-2 were lost. Capt. Apt did manage to jettison the crew capsule, but this impacted the desert floor before he could bail out.[9]

Table 1.1 summarizes the first three important American rocketplanes. The X-1B, not listed, made 27 flights and was very similar to the X-1A in both dimensions and performance. The X-1C was cancelled while still in the mockup stage. The X-1D was destroyed on its second flight after a small explosion and subsequent jettisoning from the B-50 carrier (Figs. 1.12 and 1.13).[19,20]

Fig. 1.12 "Cowboy" Joe Walker mounts his steed, the X-1A, in 1955 (courtesy NASA)

Fig. 1.13 Bell X-1E being loaded in B-29 mothership on ramp, 1955 (courtesy NASA)

NASA Lifting Bodies

From 1963 until 1975, the National Aeronautics and Space Administration, in a joint program with the US Air Force, conducted a series of flight tests of various wingless "lifting bodies" at Edwards AFB, California (Fig. 1.14). The mission of the lifting body program was to validate the concept of flying a wingless vehicle back from space and landing it precisely like an airplane. There were six different lifting bodies – the M2-F1, the M2-F2, the M2-F3, the HL-10, the X-24A, and the X-24B (Table 1.2). Some had flat tops and rounded bottoms, some had rounded top and bottom, and some had flat bottoms and rounded tops. They ranged in shape from the "flying bathtub" to the "flying flatiron." They also tried various configurations of bubble canopies, flush canopies, and different numbers of vertical stabilizers. The "M" letter designation stood for "manned," "F" stood for "flight," "HL" stood for "horizontal landing," and "X," of course, stood for "experimental." The lifting body concept had originally been developed by modifying a missile nosecone. This was essentially cut in half, and control surfaces in the form of winglets and vertical stabilizers were then added. The fact that any of these lifting bodies could be controlled as well as they were, and landed like an airplane, was nothing short of amazing.

Table 1.2 Comparison of lifting bodies

Vehicle	Length	Span	Empty wt. (lb)	Flights	Years
M2-F1	20 ft	14 ft 2 in.	1,000	77	1963–1966
M2-F2	22 ft 2 in.	9 ft 8 in.	4,620	16	1966–1967
M2-F3	22 ft 2 in.	9 ft 8 in.	5,071	27	1970–1972
HL-10	21 ft 2 in.	13 ft 7 in.	5,285	37	1966–1970
X-24A	24 ft 6 in.	11 ft 6 in.	6,360	28	1969–1971
X-24B	37 ft 6 in.	19 ft 7 in.	8,500	36	1973–1975

Fig. 1.14 Lifting bodies on lakebed, about 1970. From left to right are the X-24A, the M2-F3, and the HL-10 (courtesy NASA)

Fig. 1.15 M2-F2 lifting body landing with F-104 *Starfighter* flying chase, 1966 (courtesy NASA)

The first lifting body, designated the M2-F1, was a lightweight unpowered glider nicknamed "the flying bathtub." It was built at the Dryden Flight Research Center out of tubular steel and plywood, and completed in 1963. Using a fast Pontiac convertible, it was towed across the lakebed more than 400 times at speeds as high as 120 mph. These test-tows provided the confidence for the next step. A NASA Gooney Bird (Navy R4D, the same as the Air Force C-47 or civilian DC-3) towed the flying bathtub up to 12,000 ft and released it to glide back and land. There were 77 such aerial tow-flights. The M2-F1 proved, for very little cost, that the concept of the lifting body worked, and the green light was given to continue the research.

The M2-F2 in Fig. 1.15 was much heavier than the M2-F1. Like the Bell X-1 (and all subsequent lifting bodies), it was powered by an XLR-11 rocket engine. It was built by Northrup, and a series of drop and glide flights was conducted from beneath the same B-52 that had been used in the X-15 program. The M2-F2 had only two vertical tail fins, and was therefore prone to lateral instability. It was difficult to steer. On its 16th glide test, on May 10, 1967, pilot Bruce Peterson crashed on the dry lakebed and was severely injured. Footage of this crash was later used in the opening sequence of the mid-1970s TV series *The Six Million Dollar Man*. As a result of this crash, the craft was modified by adding a third vertical stabilizer between the other two, and redesignated the M2-F3 (Figs. 1.14 and 1.16).

The third tailfin greatly improved the controllability of the rebuilt M2-F3, and it went on to make 27 unpowered and powered flights from 1970 to 1972, using its rocket engine to reach a top speed of Mach 1.6 and a peak altitude of 71,500 ft. The M2-F3 was also equipped with a reaction jet control system similar to rocket thrusters.

Fig. 1.16 M2-F3 in-flight launch from B-52 mothership in 1971 (courtesy NASA)

Fig. 1.17 Research pilot Bill Dana and HL-10 on lakebed with B-52 flyby. The HL-10 was the fastest and highest of the lifting bodies, and was judged to be the best handling of the three original heavyweight lifting bodies – the M2-F2/F3, the HL-10, and the X-24A. Compare this photo with Fig. 1.1 (courtesy NASA)

The HL-10 (Fig. 1.17) was the highest performing of any of the lifting bodies, both in terms of altitude and speed. It was the first lifting body to fly faster than sound, in May 1969. From December 1966 until July 1970, it made 37 test flights, reaching Mach 1.86 and 90,303 ft within a 9-day period in February 1970. The HL-10 had a tall center tailfin and a flush cockpit. Its fuselage was longitudinally rounded on the bottom, and laterally curved on top.

Fig. 1.18 X-24A powered flight drop from B-52 mothership, 1970 (courtesy NASA)

The X-24A shown in Fig. 1.18 was a result of a waxing interest on the part of the Air Force to investigate lifting bodies. It was built by Martin Aircraft, had three tailfins and a bubble canopy, and was slightly larger than the Northrup vehicles. After 28 flights, the vehicle was returned to Martin and modified into the shape of a "flying flatiron" with a double-delta planiform, flat bottom, and curved top. In this configuration it was designated the X-24B (Fig. 1.19) and made an additional 36 flights.

The NASA lifting bodies together made a total of 221 flights over a period of 12 years, not including the 400-plus tow-flights made by M2-F1 with the help of the Pontiac convertible on Rogers Dry Lake.[21,22]

Today's Rocketplanes

Rocket-powered aircraft have recently made a comeback in the form of the XCOR Aerospace EZ-Rocket, the X-Racer, and the Rocket Racing League (RRL). These latest additions to the rocketplane saga are based on the Rutan Vari-EZ propeller driven airplane, with the piston engine replaced by a liquid-fueled rocket engine.

Fig. 1.19 The shape of future spaceplanes? The X-24B "flying flatiron" on lakebed, Edwards AFB, California in 1973 (courtesy NASA)

XCOR Aerospace EZ-Rocket

One of the newest rocketplanes is the liquid-fueled EZ-Rocket. It was converted into a rocketplane by a team at XCOR Aerospace in Mojave, California, from a Rutan Long-EZ. Test pilot Dick Rutan flew it for the first time on July 21, 2001. The rocketplane burns isopropyl alcohol fuel and LOX in a pressure-fed Kevlar-shielded engine, with the composite pressurized fuel tank slung beneath the fuselage. The oxidizer is stored internally in an insulated aluminum tank. Two regeneratively cooled 400-lb thrust engines power the EZ-rocket, with chamber pressures of about 350 psi and specific impulse of some 250–270 s, better than either the Bell X-1 or the Mercury Redstone. A 20-s 1,650-foot takeoff roll lifts the fully fueled EZ-rocket into the air, whereupon it can fly 10 miles in as many minutes, or climb at 6,000 ft/min. Because the pusher-prop has been removed and replaced with a rocket engine, the EZ-Rocket is aerodynamically "cleaner" than the original Long-EZ, increasing its glide range. On December 3, 2005, Dick Rutan established a point-to-point world record for the EZ-Rocket by flying 16 km in less than 10 min in a flight from the Mojave Spaceport to California City, California. This set a new world record for ground-launched rocket-powered aircraft. Recall that the NACA rocketplanes and NASA lifting bodies were all air-launched.

The EZ-Rocket has an engine burn time of 2½ min, a best rate of climb speed (V_y) of 145 knots, and a never-exceed speed (V_{ne}) of 195 knots. The maximum speed is determined by stresses on the airframe rather than by engine capability. Climbing at 145 knots gives the best performance – faster climb speeds would just create more drag on the airframe. The greatest altitude flown by the EZ-Rocket was 11,500 ft. During its 4½-year test program, the EZ-Rocket flew 26 times, including twice at Oshkosh 2002, and three times at the Countdown to the X-Prize Cup in Las Cruces, New Mexico, in 2005. These flights lasted up to 10 min, including the glide portion.

The EZ-Rocket accomplished many "firsts" in rocket aviation. In addition to the world distance record mentioned above, it was the first ground-launched rocketplane to demonstrate restart of its rocket engine in flight, the first to perform touch-and-goes, and the first to arrive at Mojave Spaceport, which it did on its last flight, December 15, 2005, returning from California City. The EZ-Rocket has a ground crew of just five and a cost per flight of only $900. A large portion of this expense is in the fuel-pressurizing system, which uses helium gas. This cost will come down when XCOR installs its in-house-developed piston-pumps in future rocketplane models, such as the X-Racer (Fig. 1.20).[23]

Fig. 1.20 Artistic impression: fleet of rocket racers over the mountains (courtesy Rocket Racing League)

XCOR X-Racer

The next, natural step forward from the EZ-Rocket is an improved rocketplane called the X-Racer, being developed by XCOR Aerospace for the RRL. Using a modified Long-EZ, the X-Racer will have a single liquid-propellant engine, the XR-4K14, burning kerosene and LOX, which are delivered to the thrust chamber by XCOR's own piston-pumps. This automatically eliminates the expensive helium-pressurized belly fuel tank used on the EZ-Rocket. Instead, the X-Racer will carry unpressurized kerosene in strake tanks built into the sides of the fuselage. The heavy LOX is carried in an insulated tank of its own, just behind the pilot and close to the center of gravity. Rather than a pair of 400-lb thrust engines, the X-Racer sports a single 1,500-lb thrust rocket, which can be shut down and electrically restarted in flight numerous times. The maximum speed – determined by the airframe, not the engine – will be 230 mph. The X-Racer will have about 3½ min of intermittent powered flight capability together with a total of some 15 min of power-off glide. This will allow the X-Racer to complete 8–9 laps around a three-dimensional aerial race track during rocket racing competition. Refueling of both the kerosene and the cryogenic oxygen will take only 5–10 min, enabling the X-Racer to get back into the race quickly.

The X-Racer and other similar rocketplanes will compete in regular events sponsored by the RRL (Fig. 1.21). This venue promises to combine the thrill of rocketry with the excitement and competition of racing. The RRL is a partner-

Fig. 1.21 Artist's depiction of Rocket Racing League Arena (courtesy Rocket Racing League)

ship of space enthusiasts and racing expertise. It was conceived by X Prize founder, Peter Diamandis, and two-time Indianapolis 500 champion team partner, Granger Whitelaw. The idea is to bring key talents together in order to "advance the technology and increase public awareness of space travel." All rocket racers will, of course, be top pilots.[25]

References

1. D. John Anderson, Jr. *Introduction to Flight*, 4th edition. McGraw-Hill, New York, 2000.
2. http://en.wikipedia.org/wiki/He_176
3. http://en.wikipedia.org/wiki/HWK_109-509
4. http://en.wikipedia.org/wiki/Hellmuth_Walter#Rocket_engines
5. http://en.wikipedia.org/wiki/Me_163
6. http://www1.dfrc.nasa.gov/Gallery/Photo/X-1/HTML/E-9.html
7. http://en.wikipedia.org/wiki/Bell_X-1
8. http://en.wikipedia.org/wiki/D-558-2
9. http://www.nasa.gov/centers/dryden/news/FactSheets/FS-079-DFRC.html
10. http://www.dfrc.nasa.gov/gallery/photo/X-1/HTML/E-9.html
11. http://en.wikipedia.org/wiki/List_of_X-1_flights
12. http://www1.dfrc.nasa.gov/Gallery/Photo/X-1A/index.html
13. http://en.wikipedia.org/wiki/List_of_X-1A_flights
14. http://aeroweb.brooklyn.cuny.edu/specs/bell/x-1e.htm
15. http://en.wikipedia.org/wiki/List_of_X-1E_flights
16. http://www.dfrc.nasa.gov/Gallery/Photo/D-558-2/index.html
17. http://www.nasa.gov/centers/dryden/news/FactSheets/FS-079-DFRC.html
18. http://www.dfrc.nasa.gov/Gallery/Photo/X-2/HTML/E-2820.html
19. http://en.wikipedia.org/wiki/List_of_X-1B_flights
20. http://en.wikipedia.org/wiki/List_of_X-1D_flights
21. http://www.nasa.gov/centers/dryden/news/FactSheets/FS-011-DFRC.html
22. http://www.dfrc.nasa.gov/gallery/photo/Fleet/HTML/E-21115.html
23. http://www.xcor.com/products/vehicles/ez-rocket.html
24. http://www.xcor.com/products/vehicles/rocket-racer.html
25. http://www.rocketracingleague.com/

Chapter 2
Why Spaceplanes?

The building of spaceplanes represents the pinnacle of human endeavor in our time. It is part of the age-old quest to understand and control the forces of nature, to strive for perfection, to scale Mount Olympus itself. It is something we humans must do. And we are doing it. As President John F. Kennedy once said, challenging America to undertake the first Lunar landing program, "We do these things not because they are easy, but because they are hard."

Within a few decades, advanced spacecraft will regularly be winging their way to and from the Moon. These Lunar spaceplanes will not only transport thousands of space tourists to and from Lunar resorts, but also serve a vital role in the space infrastructure of the future. How is this possible? For one thing, advanced spaceliners will double as space tankers, using their large capacity designs to maximum advantage. For another, they will incorporate very advanced, efficient, and versatile engines. This is just a taste of what is to come. The "Advanced Spaceplane" (Chap. 9) will feature many other superior design traits.

As with the slow and methodical development of the airplane during the twentieth century, the spaceplane will undergo a slow and methodical development of its own during the twenty-first century. This inevitable progress will be seen as a continuation in the evolution of the airplane, which was only temporarily interrupted by the introduction of the ballistic missile at the end of World War II.

What you are about to read is part aerospace history, part rocket science, and part extrapolation – based solidly on the history and the science of aerospace development. It is an honest attempt to share a vision of the future that can, and should, materialize for the benefit of all.

Space Tourism

The existing space market, which consists of the occasional satellite, scientific probe, or manned launch, will be overtaken by the space tourism industry in the coming years. The inefficiencies of these government-funded launch services will then become apparent. The ICBM-based launch architecture will eventually succumb to the undeniable superiority of the advanced spaceplane.

M.A. Bentley, *Spaceplanes: From Airport to Spaceport*,
doi:10.1007/978-0-387-76510-5_2, © Springer Science+Business Media, LLC 2009

Fig. 2.1 The prespaceplane X-38 lifting body makes a controlled gliding descent after being released from its B-52 "mothership" (courtesy NASA)

The key to this vision is the space tourist, who will make all of this possible. Beginning as early as 2009, paying passengers will begin enjoying regular "space-experience" flights in suborbital spacecraft. The vehicles that will transport these spacefaring sojourners are being built even now, as I type these words. And they are spaceplanes (Fig. 2.1).

How will the small tourist spaceplanes of today develop into the spaceliners and space tankers of tomorrow? There is no doubt they will, but just how? That is the subject of this book. One thing to keep in mind as you read along is that in space, the only way to get anywhere or transport anything is by spaceship. On Earth, there are multiple ways to travel, and they all compete with one another. Beasts of burden aside, there are various wheeled ground vehicles, some on tracks, some not. There are sea-going vessels. And there is a variety of aircraft. Each fills its own particular niche in the ecosystem of human commerce, an economic organization of amazing efficiency. This system did not grow up overnight, but has developed slowly over the course of human civilization, beginning with wooden carts and boats. In the last century we have seen the addition of aircraft to this intricate infrastructure, which are today mass-produced worldwide. The very latest additions to our planetary infrastructure have been brand new "constellations" of Earth-orbiting satellites. These include the all-important communications, navigational, and meteorological spacecraft now hovering over our Earth. As we continue our push into space, our civilization will need new ships, new ways to travel, and new infrastructure. Spaceplanes have an important role to play in the saga of space.

Fig. 2.2 The original method of flying to the Moon: Apollo 15 Saturn V Launch in July 1971 (courtesy NASA)

The days of the rocket, with its thunderous roar and billowing clouds of steam, are numbered (Fig. 2.2). Spaceplanes will inevitably replace them, in time. The space tourist will finance this change simply by engaging in space tourism. Tourism is already a multi-billion-dollar industry on Earth, and tourism in space will be no different. It will involve large sums of money being pumped into space companies, who will purchase spaceplanes for their spaceline operations. If there is money to be made in spaceplanes, you can count on spaceplanes being built in large numbers. The growth of the space tourism industry, in turn, will allow the gradual development of the spaceplanes themselves. There will be a slow and steady improvement in the performance of spaceplanes. Tourists will want to go ever higher, experience weightlessness for ever longer periods, and ride on ever sleeker craft. As bigger planes are built, more and more tourists will be able to share the experience for less and less cost, and the industry will experience a self-induced acceleration, or snowball effect.

This could happen very quickly, as millions of tourist dollars are injected into the space economy. Spaceplanes will undergo a sustained and continual improvement in size, range, and speed. Eventually the single-stage-to-orbit spaceplane will arrive, by which time space hotels will be ready to accommodate guests in Earth orbit. But the process will not stop there. Tourists will have their sights set on the Moon, and Lunar spaceplanes also will eventually appear. These events, driven by the inexorable forces of the space economy, are inevitable.

Superior Vehicle

The best qualities of the rocket and the best qualities of the airplane will be combined in a single vehicle, the advanced spaceplane. In particular, a superior airplane with advanced atmospheric utilization will be integrated with a superior air-breathing power plant, giving rise to a vastly superior hybrid vehicle. The power plant will operate as a conventional jet engine in the atmosphere, and as a rocket in space, demonstrating altitude compensation at all altitudes. Moreover, the advanced spaceplane will not only use the atmosphere but store it in liquid form for later use in space. It will be able to fill its own propellant tanks with liquefied air while airborne, instead of lifting off under the burden of this essential but extremely heavy oxidizer. The technique of using the atmosphere in clever ways to achieve the energies necessary for spaceflight will characterize the successful spaceplane (Fig. 2.3).

But why spaceplanes? Why not use advanced rockets instead? Does sending a winged vehicle, with all that dead weight, all the way to the Moon possibly make any sense? The answers to these questions require that we peer intently into the future, which is fast approaching.

Fig. 2.3 Artist's rendition of the Skylon spaceplane in orbit (courtesy Reaction Engines Limited)

Let us then take a peek at this future, imagining for a moment that we are already there. Here is what we "know." Advanced spaceplanes are superior in every respect to nonwinged space vehicles. They are reusable, safe, efficient, economical, and comfortable. Wings enable reusability and aerodynamic efficiency. Reusability increases flight frequency, leading directly to safety and economy. Comfort, reliability, safety, and economy are demanded by the space passenger transport industry, the largest market by far in space operations. Making sensible use of the preexisting aviation infrastructure, spaceplanes are now expanding that to the Moon and beyond. Spaceplanes not only depend on that infrastructure, but are a part of it. Advances in engine technology, materials, and atmospheric utilization have made the advanced spaceplane a reality in the twenty-first century. This is the essence of the advanced spaceplane.

Basic Comparisons

Let us now compare the salient features of both airplanes and rocketships, as well as future spaceplanes, to provide a clearer picture of where we've been and where we are going.

A glance at Table 2.1 shows us that the only area in which rocketships outperform airplanes is in altitude and speed. As we shall see, this is really a question of energy. In each of the other categories, airplanes have a distinct advantage over rockets, from reusability to infrastructure. A good spaceplane design will incorporate each of the good attributes of the airplane with the ability to harness great energies currently enjoyed only by the rocket. Let us look at each of these factors more closely.

Energy

Most jet aircraft have an absolute ceiling of 10 miles, or about 50,000 ft, which is about the maximum height to which jet aircraft can climb. High-performance aircraft have certainly exceeded this altitude, but ascending much above 70,000 ft

Table 2.1 Comparison of aerospace vehicles

	Airplanes	Rocketships	Spaceplanes
Operating altitude	10 miles	Unlimited	Unlimited
Operating speed	Mach 0.8	Mach 35	Mach 35
Specific energy (ft^2/s^2)	2 million	650 million	650 million
Reusable	Yes	No	Yes
Refuelable	Yes	No	Yes
Economical	Yes	No	Yes
Affordable	Yes	No	Yes
Safe	Yes	No	Yes
One-piece design	Yes	No	Yes
Simple operations	Yes	No	Yes
Good infrastructure	Yes	No	Yes

requires advanced propulsion. Spacecraft and rockets are specifically designed to reach extreme altitudes, giving them a much higher potential energy than a typical airplane. Physicists normally include the mass of an object in calculations of both potential and kinetic energies, but to compare objects – or vehicles – of unequal sizes or weights, it is convenient to divide the mass out of the equation, which yields a quantity called specific energy. This is simply the energy per unit mass of the body, and includes both the potential and kinetic energies. From these considerations, it is plain to see that any orbital or interplanetary spaceship would possess a much higher specific energy than any airplane in flight, because of the great altitudes and speeds at which it travels. At 52,800 ft, an airplane has a specific potential energy of 1.70, given in units of million feet squared per second squared. We will use these units for all of our specific energy comparisons. Spacecraft operating at altitudes of at least 100 km, or 62 miles, have specific potential energies of at least 10.6, which is 6.2 times higher than those of airplanes flying at 10 miles.

The airplane is assigned a speed of Mach 0.8, or about 600 mph, which again is typical of modern jet transports. High-performance aircraft routinely exceed this speed. Yet they never begin to approach the Earth-orbital speed of Mach 25. Rocket-powered spaceships, on the other hand, are capable of speeds greater than 25,000 mph, required to escape from Earth's gravity completely. Clearly, there is a huge disparity between the speeds of airplanes and those of rocketships. As in the case of potential energy, rockets command very high kinetic energies, compared with airplanes, because of their great speeds. An airplane flying at 600 mph, or 880 ft/s, has a specific kinetic energy of 0.387 when compared with a rocket flying at 36,000 ft/s, which has a specific kinetic energy of 648. This is 1,670 times higher.

From the above figures, the total specific energies of 600-mph airplanes and 25,000-mph rockets escaping Earth from a height of 62 miles are, respectively, 2.09 and 659. The rocket's total energy is therefore 315 times higher than the airplane's. This is why spaceflight is so difficult. It involves the harnessing of over 300 times as much energy as airplanes have to harness. On the other hand, airplanes can fly for long periods of time without running out of fuel, and never shut down their engines. Rockets run their engines for mere minutes at a time, and quickly exhaust their entire propellant supply.

It is also interesting to note the energy distribution of airplanes compared to rockets. For the airplane example given above, the energy is 19% kinetic, and 81% potential, ignoring for the moment the potential energy stored in unburned fuel. For the rocket example, the energy is more than 98% kinetic, and less than 2% potential. As the rocket escapes from Earth's gravity well, it will gradually slow down, converting its energy of speed into energy of altitude. Kinetic energy is thereby gradually converted into potential energy.

Reusability

We now come to the crux of the matter, which bears on many of the points to follow. Airplanes, with their wings and wheels, are specifically designed for repeated and frequent use. They land with all parts intact, taxi up to the ramp,

Fig. 2.4 The ballistic landing technique has certain disadvantages (courtesy NASA)

and are quickly readied for the next flight. The exchange of passengers and cargo, refueling, and other services such as routine maintenance and wing de-icing are all parts of an intricate but efficient infrastructure. This allows relatively simple, economic operations of many airplanes of all sizes and descriptions at a single airport, all under an umbrella of amazing safety and affordability.

Ever since the early Chinese fired their primitive powder rockets at the attacking Mongol hordes eight centuries ago, rocket men have been in the habit of throwing away their missiles (Fig. 2.4). On the other hand, ever since the Wright brothers made their initial heavier-than-air powered flights one century ago, pilots have been in the habit of reusing their aircraft. To this day, ballistic launch vehicles get rid of the greater bulk of their structure on every launch. This fact alone has kept the costs and hazards of rockets at a high level.

The spacelines of the future would never consider operating throwaway rockets. Given a choice between ballistic vehicles that might land anywhere in case of in-flight failure, or winged vehicles that can at least be glided and guided by a pilot, the choice is obvious (Fig. 2.5). Winged vehicles are automatically reusable, and their operation is well understood. The aviation industry has existed for over 100 years. Every aircraft that has ever flown has been designed not only for reuse, but for *daily* use. This is not the case with ballistic rockets. As a result, the safety record with aircraft is much better than with rockets, because there has been far more experience in aviation than is the case with spaceflight. For this reason alone, spaceplanes should be developed from existing airplanes by appropriate modifications. A good spaceplane should be able to operate as an efficient airplane first, and then be upgraded for spaceflight. This is far easier said than done, and will require sustained effort. Space tourism will provide the sustainable impetus to make this happen.

Fig. 2.5 The X-2 after landing on a collapsed nose wheel. With wings and pilot, both craft and occupants can fly again (courtesy NASA)

Refuelability

Arthur C. Clarke wrote about orbital refueling over 50 years ago.[1] This very sensible vision of space operations depends, of course, on rocketships retaining their fuel tanks upon reaching orbit. Yet the Space Shuttle throws away its external propellant tank on every launch. Every commercial space launcher does this. They throw away their gas tanks *and* engines in a process called "staging" to make it easier to get into space. As long as launch vehicles continue to do this, they will never be able to realize the benefits of orbital refueling. Long-term operational efficiency does not exist in ballistically launched rockets, because it has never been needed. Efficiency is routinely sacrificed for the short-sighted goal of a single launch.

Efficiency

Airplanes are economical for multiple reasons. First, their designs have continually been improved and perfected over the last century. Second, they are produced in large numbers, increasing their affordability for the private and commercial sectors. Third, they use the atmosphere in three different ways, as we will shortly see. Until recently, the only organizations that have launched and operated spaceships are national governments. Not surprisingly, spaceships have so far been neither economical nor affordable. The imminent Private Space Race is about to change all this, however. It is a very significant point, not to be overlooked, that the first reusable spaceships – the Space Shuttle and SpaceShipOne – are both spaceplanes.

The single most important factor that makes the air transport industry possible is *people* – lots of them. So it will be with space travel. A large number of space tourists will eventually drive down the cost of space access, improve the quality of space vehicles, and create affordable economies of scale for those that follow.

Spaceplanes will use the atmosphere to their advantage, instead of regarding it as a barrier to be crossed. Like all aircraft, and unlike ballistically launched rockets, spaceplanes will use the atmosphere's inherent lifting properties, its oxygen content, and its propulsive potential. Ballistic missiles and rockets do none of these. Hybrid engines incorporating elements of the turbofan and the aerospike can be designed into the successful spaceplane. Such "spinning aerospikes" will therefore be able to use bypass air in the atmosphere as propellant, and employ automatic altitude compensation, ensuring peak efficiency all the way into orbit. High-altitude oxidizer transfer will be another technique spaceplanes will use to reach low Earth orbit. Spaceplanes need not take off with heavy loads of both fuel and oxidizer onboard. Because liquid hydrogen – the most energetic rocket fuel available – has a very low specific weight, spaceplanes may take off fully fueled with LH_2 but with empty oxidizer tanks. At some 50,000-ft altitude, the heavy oxidizer – liquid air, hydrogen peroxide, or liquid oxygen – will be transferred from an aerial tanker to the spaceplane, which will then immediately accelerate into orbit from this altitude.

Once in orbit, spaceplanes will replenish both their fuel and oxidizer tanks from an orbiting propellant depot supplied by regular spaceplane sorties. With fully replenished tanks, and with a required delta-V of only two additional miles per second on top of the 5 mps already attained for low Earth orbit, the spaceplane will have the capability of leaving Earth orbit. The spaceplane may now become a Lunar shuttle, landing on the Moon with relatively low-thrust ventral rockets. Returning to Earth after a quick Lunar turnaround, the Moonplane will make a direct reentry into Earth's atmosphere at 25,000 mph and land at any spaceport, to be refueled and reflown in a few hours.

Safety

Today's spaceships are probably as safe as they can be, given the technology available. Every time a passenger boards an airplane, the chances of perishing during the flight are about one in 5 million. The risk levels are therefore relatively low. In the case of the Space Shuttle, however, nearly one in sixty flights has killed the entire crew. And yet the Shuttle is considered safe enough to operate, even though the statistical risk of a catastrophic failure is almost a hundred thousand times the level of a commercial airliner. But why are today's spaceships so much more dangerous than today's aircraft? The answer has to do with how they leave the ground (Fig. 2.6).

Fig. 2.6 Ballistic launch of Space Shuttle *Columbia* on April 12, 1981 (courtesy NASA)

Most spacecraft today, including the Space Shuttle, are launched as ballistic missiles, resulting in very high risk factors. This point is fundamental in understanding the huge difference in safety between rockets and airplanes. Because conventional spaceships are typically launched by nonreusable ballistic missiles, they have undergone far fewer test flights than airplanes, all of which have been through extensive flight test programs. Spaceflight is very difficult, presenting a much greater challenge than the relatively benign speeds and low altitudes of aerodynamic flight. Hence space launches have occurred far less often than flights of airplanes. Another factor exacerbating this situation is an immature market. Since 1957, the only customers for Earth's space launch industry have been a few hundred satellites and, of course, government space programs. Space tourism is about to change all that.

Simplicity

Rockets are perhaps unique in the field of engineering in being specifically designed to come apart during normal operation (Fig. 2.7). The Saturn V Moon rocket, for example, broke into no less than eight pieces on every flight. Only one of those pieces, and one of the smallest, returned to Earth. This modular flight

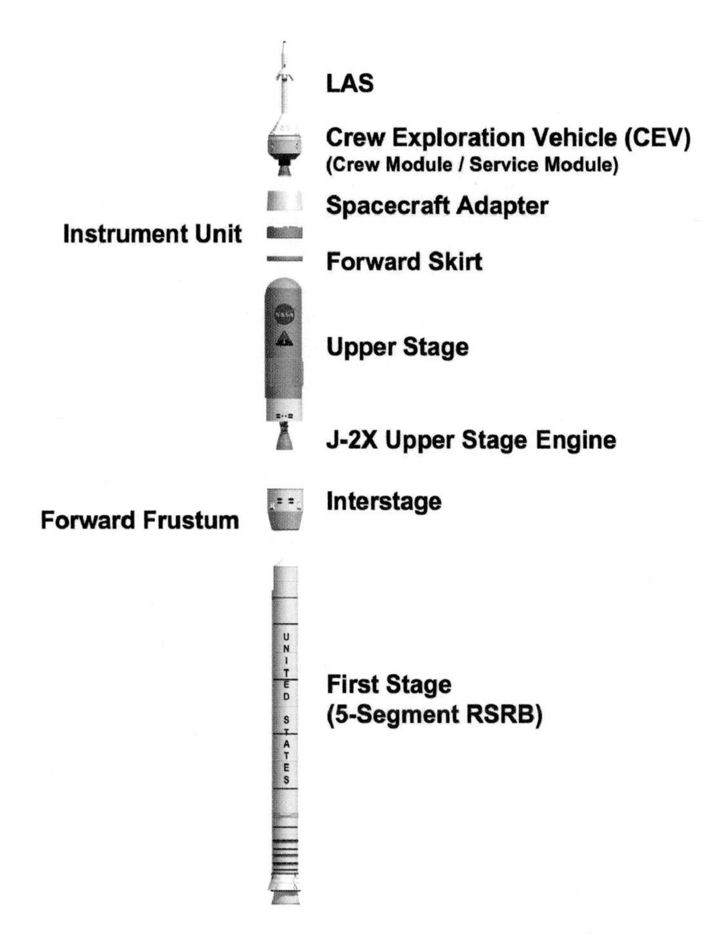

Fig. 2.7 Diagram of the planned Ares I crew rocket, showing typical modular design. The only reusable parts are the solid rocket boosters and the conical crew module (courtesy NASA)

architecture worked satisfactorily, but it was exceedingly wasteful in terms of discarded hardware as well as cost. The Space Shuttle does a little better, breaking into only four pieces, but still ludicrously throwing away the most operationally important component, its propellant tank. Spaceplanes, by contrast, will specifically be designed to retain all parts, refuel almost anywhere, and return for new missions.

Spaceplanes will operate from runways, much like ordinary airplanes. They will not require extensive launch complexes, mobile service structures, or armies of launch personnel. Furthermore, they will be designed to be landed, cooled down, refueled, and reflown within hours, rather than months or years as with the Space Shuttle. Simple airplanelike operations will characterize the successful spaceplane of the future.

Operational Superiority

The key to a successful spaceplane design concept is to identify whatever airplanes do better than rockets, as well as whatever rockets do better than airplanes, and incorporate these features into a single self-contained vehicle. That, in a nutshell, is the challenge of the spaceplane. For example, airplanes use the atmosphere in no less than three distinct ways: (1) aerodynamic lift, (2) oxidizer to sustain combustion of fuel, and (3) propellant or "working mass." Rockets, by contrast, use the atmosphere for none of these, instead carrying their own onboard propellant-oxidizers, and replacing aerodynamic lift by brute rocket thrust. These factors do allow the rocket to operate in the vacuum of space, however, where ordinary aircraft would be completely helpless and could never reach in any case.

The Space Shuttle solves only half the spaceplane equation, using its wings to land on Earth, but resorts to ballistic methods to enter space. This means that the Shuttle's wings and vertical stabilizer are dead weight during launch. It is still a step in the right direction, but advanced spaceplanes will go a step further. Utilizing the horizontal takeoff and horizontal landing approach, spaceplanes will cater much more readily to their passengers, while avoiding the pitfalls that have plagued the Shuttle program. They will operate routinely from aerospaceports around the globe. Horizontal takeoff means that spaceplanes will be bothered neither by the kind of falling debris that eventually doomed the *Columbia*, nor by hot gases escaping from strap-on solid rocket boosters, which resulted in the *Challenger* explosion. We will report on each of these tragedies later in the book, with a view to applying the lessons they teach us to future spaceplane design.

The main argument against the development of single-stage-to-orbit spaceplanes is the belief that they would not be able to carry enough propellants to reach orbit. Indeed, the Shuttle's space transportation system carries the bulk of its propellants in two huge solid rocket boosters and an enormous liquid propellant tank, both external to the delta-winged orbiter itself. And yet, the main problem with the Shuttle is that it is launched vertically. The spaceplanes of the future will all be launched horizontally, and will therefore not require nearly the propellant capacity of the Space Shuttle, because of greatly improved atmospheric utilization. In addition, spaceplanes will be far more versatile than either aircraft or present-day spacecraft have been. Most spaceplanes, at least in the short term, will be specifically designed to carry passengers and crew rather than freight. The Space Shuttle, by contrast, was designed as a heavy-lift "space truck" able to lift up to 65,000 lb of cargo into low Earth orbit.

Versatility and Infrastructure

As better and better spaceplanes take their shape from the drawing boards, and eventually fling themselves into the heavens, they will begin to prove their versatility. One area in which the advanced spaceplane of the not-so-distant future will shine is in the

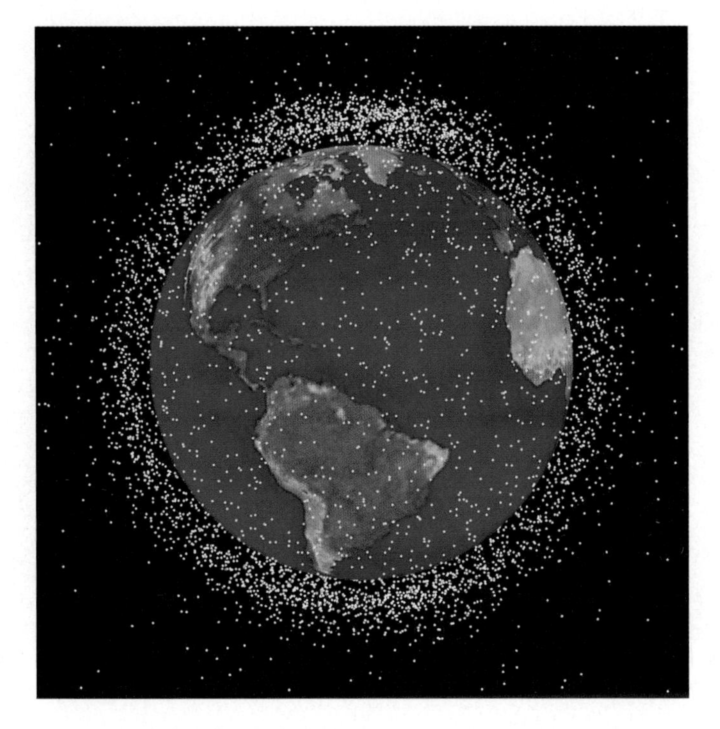

Fig. 2.8 Low Earth orbit "Beehive" of orbital debris caused by ballistic rocket refuse (courtesy NASA)

area of propellant shipping. Not only will spaceplanes rely on propellant depots in space, but they will very likely be the vessels to keep those depots supplied.

Over the course of the last century, a highly developed infrastructure has grown up to support the efficient operation of all aircraft. There are conveniently placed fuel stations, repair shops, and readily available spare parts. When maintenance or repairs are required, the shops, parts, and mechanics are on hand to fill a vital need. The first spaceplanes will make use of existing aviation infrastructure, and gradually expand it into near-Earth space. Propellant caches will be placed in Earth orbit, on the Moon, and on other planets, in order to support a mature space transportation system in the twenty-first century. Aerial tankers may also service spaceplanes just before they "light off" for orbit. And, of course, highly trained technicians, workshops, and spare parts will be strategically placed at suitable sites.

Ballistic Refuse

Throwaway launch vehicles, upper stages, and exploding satellites have littered near-Earth space with thousands of pieces of orbital debris, as seen in Figs. 2.8 and 2.9. This refuse is not only wasteful but extremely hazardous to spacecraft operating

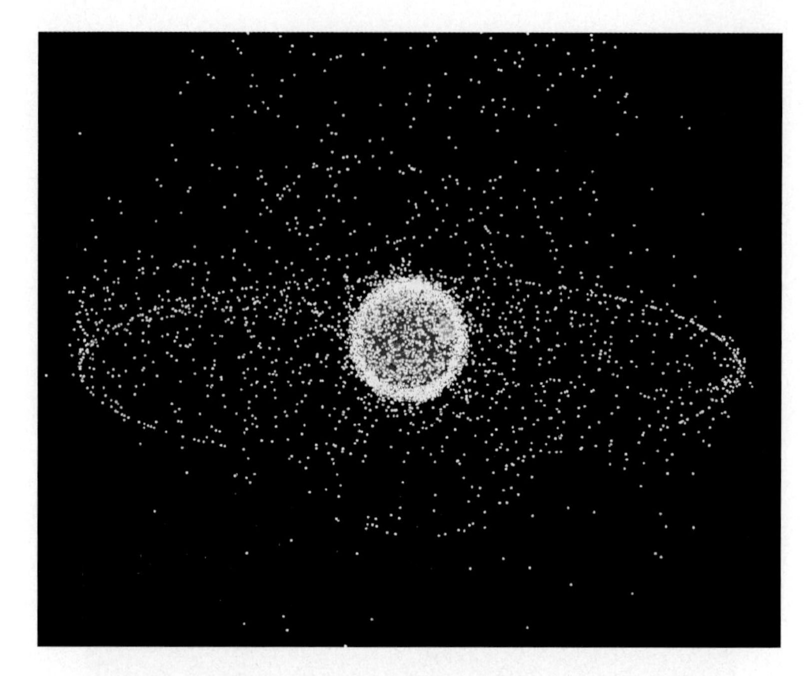

Fig. 2.9 Geosynchronous orbit debris pattern (courtesy NASA)

in the vicinity. The US Space Surveillance Network routinely tracks all debris larger than 10 cm in size – about 11,000 pieces – to ensure that orbiting satellites and manned vehicles are not endangered. Besides this "large" orbital debris it is estimated that there are more than 100,000 "little" pieces between 1 and 10 cm in size, and millions of particles smaller than this. The fully reusable spaceplanes of the future will not contribute to this diabolical danger, because they will not shed, or shred, pieces of themselves into space as they leave the atmosphere. Near-Earth space is defined as the region within 2,000 km of the surface. This shell is accumulating a large amount of rocket rubbish with each launch. The solution to the debris problem – and it does not take a rocket scientist to figure this out – is to stop adding to it. Natural orbital decay of these particles will eventually clear the near-Earth environment. And spaceplanes, with their self-contained, nonmodular designs, will greatly help to alleviate the problem.[2]

Conclusion

The technology is on the horizon to realize this vision within a few years. The various components simply need to be put together in the right way to create a working system. The ingredients are regular runways, wings, wheels, a lifting atmosphere, tanker aircraft, reusable space tankers, orbital fuel depots, air-breathing turborockets, pilots, and passengers. Most of these items are already in place.

In the chapters that follow, we will take a much closer look at some of these exciting possibilities. If you are reading this book, then you could help make it happen.

References

1. Arthur C. Clarke, *Interplanetary Flight*. Harper & Row, New York, 1950; Berkley, New York, 1985. Chap. 4.
2. http://www.orbitaldebris.jsc.nasa.gov

Chapter 3
Rocket Science: Wings Added

The date is sometime in January 1945. The place, the North Sea German rocket installation known as Peenemünde. The rocket, the dreaded V-2, is about to be launched, but something is strange here. This V-2 has large swept-back wings. The countdown starts... Zehn!... Neun!... Acht!... Sieben!... Sechs!... Fünf!... Vier!... Drei!... Zwei!... Eins!... Feuer! The vertical winged A-4b, as this new creation has been christened, thunders to life and roars off the launchpad. Soon it has disappeared from view, but is tracked with long-range instruments. The winged rocket flies effortlessly through the speed of sound, coasts upward to the peak of its trajectory, and arcs over into a high-speed dive. A slight buffet arises as one of the wings begins to work itself loose. Then it suddenly rips off at Mach 4. The first supersonic rocketplane has just been born – and killed.

Ballistic missiles are structurally simpler than any winged vehicle, with the requirement to handle only longitudinal loads rather than transverse stresses as well. Toward the end of World War II, German engineers affixed wings to their V-2 with a goal of increasing its range. In the first launch attempt of the modified missile, the rocket failed for reasons having nothing to do with its newly attached wings. In the second attempt, the vehicle worked as intended, and the rocket attained gliding flight for a time before aerodynamic forces ripped one of the wings off. But the experiment proved that a ballistically launched rocket could become a winged glider, presaging the Space Shuttle by almost 40 years (Fig. 3.1).

The German experiments with winged versions of the V-2 show how difficult it can be to marry the two independent technologies of aeronautics and astronautics. For although they both involve flying machines, the similarities end abruptly at that point. Aeronautics has always immersed itself in a relatively benign atmosphere full of oxygen, lift, and propulsive potential. Rocketry, on the other hand, has always been concerned with getting above the atmosphere and into the vacuum of space. Aircraft rely on the atmosphere in many ways. Rockets shun it completely, preferring the bold independence of the impulsive dash into orbit.

The development of the spaceplane insists on the marriage of the two apparently unrelated sciences of aeronautics and astronautics, a daunting task. For although aircraft tend to take off in a horizontal attitude, space vehicles are typically lofted from their launch pads in a vertical fashion. Aircraft and rockets are therefore built to handle very different stresses and loads. Rockets are simple vertical tubes, filled

M.A. Bentley, *Spaceplanes: From Airport to Spaceport*,
doi:10.1007/978-0-387-76510-5_3, © Springer Science+Business Media, LLC 2009

Fig. 3.1 Bomarc Missile making vertical launch. Wings extend the missile's range (courtesy USAF)

with fuel and oxidizer, constructed specifically to endure longitudinal loads, whereas aircraft must support transverse loads through their wings and landing gear. These factors clearly distinguish the design of rockets from aircraft, illustrating just one of the challenges in melding these two approaches in a single vehicle.

Let us now take a look first at airplanes, then at rockets, to gain some appreciation for each of them separately, before we examine the fusion of these two technologies in the spaceplane.

The Airplane

Aircraft operate continuously within the Earth's atmosphere, relying on this gaseous fluid in no less than three distinct ways. First, the atmosphere is essential in generating a lift over the wings to counter the downward pull of gravity. The difference in air pressure, above and below the wing, accounts for the upward force of lift. Second, air contains a critical ingredient, namely, oxygen, which sustains combustion inside any engine. Airplanes therefore need not carry their own oxidizer any more

Fig. 3.2 Spaceplanes operating in the atmosphere depend on the same aerodynamic principles that govern the flight of these research aircraft (courtesy NASA)

than you would bother with an oxygen bottle for the family car. Third, and this point is easy to overlook, the atmosphere provides the "working fluid" for the thrust generated by the aircraft engine. Again, the aircraft has no need to carry its own propellant, since it makes use of the air all around it to propel it forward. Propeller-driven aircraft accelerate the oncoming air past an airscrew, which provides the required forward thrust. In the case of the jet engine, air is sucked into the front end, compressed, combusted with fuel in the turbine, and shot out the rear end. In a turbofan, bypass air is brought in and shunted around the engine core itself, and used as the major portion of the total thrust in the exhaust stream.

Four basic forces are involved in the operation of any aircraft (Fig. 3.2). These are lift, weight, thrust, and drag. They occur in oppositely directed pairs, so that aerodynamic lift counters weight, and engine thrust overcomes air resistance or drag. Lift depends in a complex way on airspeed, air density, and a coefficient of lift unique to each wing. It is computed by taking the difference in air pressure above and below a wing, and multiplying that number by the wing area. Large wings, therefore, generate more lift than small wings do. The gravity force, or weight, pulls the aircraft always toward the center of Earth, so lift must continually counteract gravity while the airplane is in flight. In a similar manner, thrust and drag forces act in mutually opposite directions on an aircraft in flight. The thrust vector is pointed forward, while the drag vector is directed backward with the relative wind. These two forces must also balance for stable flight conditions to prevail. If these two forces are not equal, then the aircraft will either speed up or slow down.

Fig. 3.3 Space Shuttle *Atlantis* about to touch down at Edwards AFB, California, upon completion of the STS-66 mission (courtesy NASA)

Should the all-important airspeed drop below a certain critical value, the wings will stop generating any lift at all, and the airplane will stall. This has nothing to do with the engines. A stalled airplane is simply one whose wings have stopped producing lift. All aircraft must continue to develop lift over their wings until they land, or lose control and plummet to Earth. This applies to the Space Shuttle the same way as it does to any aircraft, more so because the Shuttle is an unpowered glider on every landing (Fig. 3.3). How, then, can the Space Shuttle fly at all? Where does its thrust come from?

Just before landing, the Shuttle actually depends on the force of gravity for its forward thrust. Maintaining its nose-down attitude until just before the touchdown flair, the gravity vector can be divided into two mutually orthogonal vectors, one in line with the direction of travel, and the other normal (perpendicular) to the belly. The vector component normal to the belly is countered by the lift generated by the Shuttle's wing and body, and the vector component in line with the direction of travel effectively serves as the thrust, which is countered by the aerodynamic drag.

The basic airplane maintains positive control during flight by using aerodynamic control surfaces. The three axes of control are pitch, roll, and yaw. Pitch determines the angle of attack, and is controlled by elevators mounted on the horizontal stabilizer or by forward-mounted canards. These small control surfaces move up and down together, allowing the pilot to pitch the nose up or down as desired. Roll determines the degrees of bank, and generally allows a pilot to keep the wings level during cross-country flights, or execute coordinated turns. Stunt pilots may perform barrel rolls while maintaining a constant heading. Ailerons mounted on each wing deflect in opposite directions simultaneously, causing the aircraft to roll about a longitudinal axis. Yaw determines aircraft heading and is controlled by means of a rudder mounted on the vertical stabilizer or "tailfin." An airplane's rudder works exactly the same way as a boat rudder, and serves the same purpose. To make a coordinated turn, the skilled pilot uses both roll and yaw control at the same time.

The Rocket

Just as the airplane relies on laws of flight physics that are fundamentally different from the laws governing the buoyant flight of balloons and airships, so the rocket makes use of unique physical laws in its own operation. These laws were first written down by the English physicist Sir Isaac Newton, and are basic to the flight of all spacecraft:

1. An object in motion or at rest remains so unless influenced by an external force.
2. A force imposed on an object causes an acceleration proportional to its mass.
3. For every force or action, there is a counter-force or reaction.

We will examine these laws in reverse order, since this permits a more logical discussion. The third law explains how rockets work, and especially why they work in the vacuum of space. In a nutshell, the velocity of the exhaust gases exiting the rocket engine builds up a momentum, causing the rocket to accelerate in the opposite direction. The greater the momentum – mass times velocity – of the rocket exhaust, the faster the rocket-powered vehicle will go. A toy balloon released after inflation is a good example of the rocket reaction. When a rocket operates inside Earth's (or any planet's) atmosphere, the ambient pressure conditions affect the operation of the engine, usually resulting in less-than-efficient operation. Rockets are therefore happiest in the vacuum of space, where they can operate at optimum performance.

The second law tells rocket scientists how fast a spacecraft or launch vehicle will speed up when the engines are fired. A heavy vehicle, fully fueled, will obviously accelerate very slowly, while a lighter one will accelerate much more rapidly under similar thrust. The less mass a rocket has to accelerate, the easier it is to speed that rocket up. And speed is everything in spaceflight. This naturally intuitive concept explains why all launch vehicles shed pieces of themselves on the way into orbit. For every stage that is dropped off, the remaining rocket has a significantly easier job getting the rest of the vehicle up to the required speed.

The first law is what allows a spaceship to coast, in the absence of air drag, almost all the way from its launch point to its ultimate destination without firing its rocket engines (Fig. 3.4). It is also what allows orbiting spaceships, stations, satellites, and probes to freely operate for extended periods. All orbits may be described in terms of free-fall trajectories.

Rockets do not need to push against anything for their operation. In fact, they work better in the vacuum of space than they do in the atmosphere, as described above. Just how do rockets work, though? Why do they always seem to lift off vertically?

Rockets are the ultimate in controlled combustion heat engines. They work by taking the potential energy stored in a fuel, burning that with an oxidizer in a specific

Fig. 3.4 Artist's concept of Orion spacecraft with attached Lunar Surface Access Module approaching Luna (courtesy NASA)

way, and turning that energy into the potential energy of altitude and the kinetic energy of high velocity. This explains how the Space Shuttle, for example, starts out with a huge external tank filled with more than 1½ million pounds of liquid propellants and two enormous solid propellant boosters, and 8½ min later winds up in a low Earth orbit at a speed of 17,500 mph, having burned virtually all of its propellant.

Fuel and oxidizer are injected into the thrust chamber of a rocket engine, either through the operation of high-pressure turbopumps or by means of an inert gas pressurization system. The high pressures are necessary because of the voracious appetites of rocket engines. As combustion takes place, the hot gases can escape in one direction only, which is out the rear through a specially designed supersonic nozzle. The nozzle is designed to accelerate the exhaust gases with maximum efficiency, because it is the acceleration of the gases that imparts the required momentum to the vehicle itself in the opposite direction.

Unlike airplanes, most rockets depart vertically from the ground. Aside from the obvious reason that the destination is straight up, there are other considerations that compel this approach. In contrast to other vehicles, rockets carry everything they will need with them. This includes the oxygen they will need to sustain combustion of the fuel in their engines. The onboard propellants – fuel and oxidizer – serve the dual purposes of providing the energy and the working mass for the thrust required. Because there is no air in space, all the working mass must be brought along, and it is convenient that the fuel and oxidizer fill this role. Therefore, launch vehicles have no need of the atmosphere at all. In fact, it represents a barrier to be crossed as quickly as possible. Too much speed in the atmosphere can result in dynamic pressures that can tear a vehicle apart. Drag losses also increase the required "delta-V" to reach orbit. So the quicker a rocket can punch through the atmosphere – before it has built up too much speed – and reach the rocket's real element of space, the better. Further, vertical launch permits a rocket to be designed in a structurally simple way, because the acceleration loads are mainly longitudinal – along the length of the stack. Finally, rockets have no requirement for landing gear or wings, because they never return to Earth. This further simplifies the design of these vertical, nonreusable tubes. These factors make rockets both lightweight and structurally sound for their one-time missions. For all these reasons, rockets have always been launched straight up.

The Rocket Equation

To fully appreciate any rocket or rocket-powered spacecraft, a good understanding of one of the fundamental equations of rocket science is essential. The equation relates several basic parameters of spaceflight in one simple formula.

$$\Delta V = c \ln (M/m)$$

The first item in the formula is the all-important ΔV, which is the theoretical change in velocity needed for a particular space mission. This value includes not only the velocity required, for example, to reach low Earth orbit, or go to the Moon, but actually incorporates other factors, such as drag and gravity losses, landing maneuvers, and midcourse corrections. For example, a typical ΔV for getting into low Earth orbit is 30,000 ft/s, which includes about 4,300 ft/s to cover atmospheric drag and gravity losses.

The second factor in the rocket equation is exhaust velocity, c, which is the speed of the exhaust gases as they pass the exit plane of the rocket nozzle. Once past this point, the expanding gases have done all the work they can do in accelerating the rocket. The exhaust velocity depends on the propellant's specific impulse, I_{sp}, which is the thrust divided by the weight of propellant burned per second. The higher the total thrust per unit weight flow rate of propellant, the higher the specific impulse will be. And the lighter weight the propellant, the higher the exhaust velocity will be. This is why Konstantin Tsiolkovskiy wrote about the spaceflight benefits of hydrogen fuel over a hundred years ago. There is more information on exhaust velocity and specific impulse in the appendix.

Finally, we come to the natural logarithm of the mass ratio, M/m. The mass ratio is simply the initial fully fueled mass M of the vehicle divided by the final mass m, after burning off all propellants. There are two ways to increase this ratio. The first is to maximize the initial weight by using denser propellants, and in fact this technique is used in the first stages of many rockets. The second is to minimize the final weight by making the empty structure of the vehicle as light as possible. One of the greatest challenges in spaceplane design is to get the mass ratio as high as possible in a single vehicle. The easiest method of increasing mass ratio is through a technique known as staging – used by virtually all rockets. The natural logarithm is an exponential curve with a horizontal asymptote, meaning that the higher the mass ratio becomes, the less incremental benefit that accrues. For example, doubling the mass ratio increases the ΔV, but it does not quite double it, because the logarithmic curve levels out.

Staging

Rockets are typically composed of two or more stages, or segments, designed to fall off as they use up their propellant. This greatly reduces the weight of the remaining vehicle to be boosted into orbit, and is a highly effective means of achieving orbital speed. The rocket equation is applied as many times as there are stages on the vehicle, with the final mass, m, being equal each time to the weight of the stack just *before* the expended stage drops off. As each successive stage ignites, the new initial mass, M, is equal to the sum of the remaining fully fueled stages and the spacecraft atop the rocket. The great advantage in using multiple stages is an increase in effective mass ratio. This results in a greater payload weight being delivered to an intended orbit, either around Earth or for transfer to an objective beyond. The increased payload capability can be traded for speed, using a given launch vehicle, by placing a much lighter payload on top of the booster stages. This was done with the Pluto-bound New Horizons spacecraft, which had to be made as light as possible in order to reach the environs of Pluto within a decade. New Horizons took only 10 h to pass the orbit of the Moon, compared to 3 days for the Apollo missions.

Rockets have good reasons to use multiple stages to accomplish their typically one-time missions. These reasons, however, do not apply to the successful spaceplanes of the future, because the respective missions of ballistic rockets and spaceplanes will be fundamentally dissimilar. The most important difference between spaceplanes and rockets is that spaceplanes will be reusable.

Now that we have taken a quick look at airplanes and rockets, let us examine the vehicles of the future, the spaceplanes that will open up space to the traveling public.

The Challenge of the Spaceplane

Single-stage-to-orbit spaceplanes represent the ultimate design challenge in the field of aerospace engineering. The dual aims of maximizing both mass ratio – required for getting into orbit with a single vehicle – and payload capacity – the reason for going in the first place – directly conflict with one another. Although it is theoretically possible to achieve orbit with a single-stage vehicle, most designs result in a very small payload capability. By the time such a vehicle reaches low Earth orbit, the spacecraft is essentially a floating gas tank with a ridiculously tiny passenger cabin or cargo bay. Something like 90% of the vehicle is completely empty. This is the basic reason we are still stuck on vertical-staged rockets. Can something be done about this?

New vehicles require bold, new thinking. The first thing to realize is that disadvantages can be turned into advantages. As a simple illustration of this, consider the potential of a floating winged gas tank that just happened to be refueled in orbit. Never mind for now where the gas comes from. Just pretend for the moment that you are the pilot of a spaceplane that has just reached LEO and you are out of propellant. You only have enough fuel to place yourself in an orbit that will allow rendezvous with a supply depot. Once fully replenished with fuel and oxidizer, what are your mission capabilities? The answer may surprise you.

Remember, the spaceplane you are piloting has a ΔV capability of 5 miles/s, fully fueled. You have used every ounce of this capacity to reach orbit, using a few other tricks to reduce propellant consumption, and minimize drag and gravity losses. So now you are sitting in orbit, you have just undocked from the depot, and you have a full gas tank. Your supply of onboard stocks is too small to attempt an interplanetary flight, since you would likely run out of food, water, and breathing air during the months-long transfer ellipse. But with full tanks, you are just itching to go *somewhere*. Fortunately, there is a large destination nearby: the Moon. It is only another 2 miles/s to achieve escape velocity from LEO, and returning to Earth is no problem, because you have wings. Getting in and out of Lunar orbit will not take much fuel, because the gravity field of the Moon is weak and you will have slowed down significantly during the uphill coast to the Moon. So you light off your engines once again, and you are on your way. Within 3 days, you, your crew, and your passengers are gazing down on the pock-marked surface of Luna from the cozy cabin of your sleek spaceplane. But wait a minute! You just remembered something important.

Your gas tanks still have the equivalent of almost 3 miles/s ΔV in fuel and oxidizer, and you will not need nearly that much to get back to Earth. It is downhill almost all the way, and the atmosphere is your free air-brake. Just then, you have an idea. The surface base has informed you that they are running low on fuel, and your passengers have been pestering you for a Moonwalk. But you cannot land the spaceplane on the Moon, because it is designed for landing only on Earth. You have no intention of making a tricky tail landing, because you do not want to tip over. So you give the crew at the base a call.

> "Tycho One, Spaceplane Yeager."
> "Go ahead, Yeager."
> "We've got a proposition for you. We just happen to have around half a tank of gas that we don't really need for the coast back. Seems we also have some tourists who've been itchin' for a Moonwalk, but they would need a lift to the surface and back. Could you maybe help us out?"
> "Roger that, we'll rendezvous with you in two hours. Glad to help out." With that the deal is done. Two hours later, the spidery Moon shuttle docks with *Yeager*, hatches and hoses are sealed and safed, and passengers and propellant are promptly transferred. Another 24 hours at the Moon, space tourists back aboard, trans-Earth injection burn complete, and you're coasting home in the fastest plane ever built.

In a few days you will be entering the atmosphere at a searing 25,000 mph, make a few skip-entries to slow down and stay cool, and make your final entry and landing a week after you left the spaceport.

This little story serves to illustrate how versatile spaceplanes will be in transporting both passengers and propellant to the Moon. Because of the fact that any single-stage-to-orbit spaceplane will have large propellant tanks, it is automatically well suited for space tanker duty. In this way, the huge tanks can be useful not only in getting the vehicle into orbit, but also in supplying various bases and depots with precious propellant. Let us examine this idea a little more closely.

What if water is found on the Moon? Of course, it would exist as ice, probably in the environs of the Lunar poles. If this ice could be melted and electrolyzed into its constituent elements of hydrogen and oxygen, the Moon itself could supply large quantities of propellant for future space infrastructure. This completely changes the picture. The story would now have its heroes picking up Lunar propellants and delivering them to the Earth-orbiting depot, rather than the other way around. As far as getting to the Moon is concerned, the spaceplane would need only a fraction of the fuel that it needed to get into orbit. So it would fill its tanks to about two fifths of their capacity, fly to the Moon with a much lighter load, and then return with a full load of water, not rocket propellants. This is because water takes up less volume than the same material in the form of liquid hydrogen and liquid oxygen. When it comes to utilizing propellants, it is mass that counts. Once the Lunar water is delivered to the LEO propellant depot, it can be electrolyzed into rocket propellants using solar energy. This would be done just before the next scheduled arrival of a thirsty spaceship.

Fig. 3.5 Fast mother – baby spaceplanes have a lot to learn from swift mamas such as the SR-71 *Blackbird* (courtesy NASA)

That is a peek at the future. But just how will these advanced spaceplanes be conceived, how will they be developed, and when will they be born? To answer these questions, we need to fully appreciate the baby's parents, its "mother" the airplane and its "father" the rocket.

A Hybrid Is Born

What makes an airplane so successful? As we saw in the last chapter, it is its inherent efficiency as both an aerodynamic and economic vehicle that accounts for this. Airplanes are mass-produced, plugged into an existing infrastructure, and have long lines of paying passengers clamoring to climb aboard.

The airplane has a lot going for it. Not only does it have people who want to pay for rides, but it has an atmosphere willing to provide free oxygen, glad to exchange aerodynamic lift for a little atmospheric drag, and serve freely as the all-important working mass (Fig. 3.5).

The rocket has a lot going for it as well. It has large energies at its command, and has the capability of reaching the equilibrium state of orbital free-fall, in return for a brief impulse from its powerful engines. Once it reaches orbit, it can shut its engines down completely, and coast around the world for free, essentially forever.

The free components of spaceplane operations, then, will include oxygen from the air, working mass from the atmosphere, and the free-fall of orbit. The prices that must be paid are drag in exchange for lift, and propellants in exchange for speed.

Fig. 3.6 Baby spaceplane X-38 cradles under the wing of its "grandmother" B-52 (courtesy NASA)

Now comes the challenge: combining the rocket with the airplane. They must be integrated successfully, and give birth to a hybrid vehicle with exceptional qualities superior to its forebears (Fig. 3.6). This baby will have far more potential than the dwarf rocketplane X-15, and be far less clumsy than the awkward Space Shuttle. This vehicle, when fully grown, will be able to stand on its own legs (not its tail), and land on its own feet. It will be far better adapted to its environments than either of its parents, flying in the air faster than its airplane "mother," and flying through space farther than its rocket "father." A precious spaceplane is born.

This has been a very short survey of some of the theoretical challenges involved in building spaceplanes. In the remainder of the book, we will look at these issues more in-depth in an attempt to gain some deeper insight.

Chapter 4
Missiles and Modules

The advent of the spaceplane insists on new thinking about the way we access space. Instead of launching vertically from a launchpad, we must accustom ourselves to vehicles that take off from runways, like any plane. Instead of casting off pieces of launch vehicle on the way into space, or settling for a few reusable components, we must insist on complete reusability. Instead of months spent between missions, we must think of routine operations that allow spaceplanes to be flown several times *a day*. Instead of Moon modules, we must think in terms of truly versatile spaceships. Instead of being the domain of the select few, we must accept the possibility that there is room in space for the masses. The spaceplane is the key that will help us realize these aspirations.

Let us contrast today's ballistic rockets and vertical vehicles to tomorrow's space vessels. By the time you finish this chapter, you should have a good understanding of the strengths, weaknesses, and potential of most types of space launch vehicles, including the spaceplanes of the near and distant future. We will ask some tough questions, and seek honest answers. Questions such as, will spaceplanes someday completely replace conventional rockets? What is the Achilles' heel of all ballistic missiles? Why will spaceplanes be safer than vertical launch vehicles? When will these types of aerospace vehicles appear? What is the greatest strength of the spaceplane? What is its greatest weakness?

Historical Rocketry

Rockets trace their ancestry back about 800 years to the Chinese, who used them against invading Mongol hordes in the year 1232. Not much is known about these early Asian devices; they are described simply as "arrows of flying fire."[1] Centuries passed with little improvement in the design of these early black powder rockets, but those versed in the "black art" kept the science alive through the ensuing centuries. By the early nineteenth century Sir William Congreve of the British Army made several notable improvements, including larger size and range, improved propellant-packing techniques, sheet-metal casings, and portable center-guide sticks.[2] Congreve rockets were used at the battle of Fort McHenry during the War

M.A. Bentley, *Spaceplanes: From Airport to Spaceport*,
doi:10.1007/978-0-387-76510-5_4,

of 1812, and are immortalized as "the rockets' red glare" in the national anthem of the United States. William Hale later made further improvements, notably the addition of exhaust vanes, which gave the rockets a stabilizing spin. With the advent of longer range artillery shells and better aiming methods, rockets almost faded entirely from the scene until the early twentieth century (Fig. 4.1).

It was not until 1926 that a pioneering American physicist, Dr. Robert Hutchings Goddard, built and flew the first liquid-fueled rocket (Fig. 1.2). The advantage of all liquid-fueled rockets is that they can be throttled. They can be turned off or throttled down for various reasons, giving them a distinct advantage over solid rockets. They also tend to be more efficient, especially at high altitudes. Within 20 years, Walter Dornberger and his team of German rocket scientists, including Dr. Wernher von Braun, had developed a series of liquid-propellant rockets culminating in the famous V-2 of World War II. All subsequent liquid-fueled rockets can trace their ancestry back to the V-2 (Fig. 4.2).

Liquid-propellant rocket engines are marvels of engineering. They may use a range of cryogenic or storable, inert or hypergolic, toxic or nontoxic chemicals in their operation. Typically high pressures are required as two liquid propellants – a fuel and an oxidizer – are admitted into the combustion chamber. The high pressures are delivered either by pressurized propellant tanks or by high-pressure turbopumps, depending on the design.

Fig. 4.1 Small 3.25-in. ballistic missile rapidly accelerating skyward from NACA's Wallops Island facility on June 27, 1945 (courtesy NASA)

Fig. 4.2 Dr. Wernher von Braun, designer of the German V-2, Mercury Redstone, and Saturn family of launch vehicles. Saturn IB stands in the background (courtesy NASA)

Types of Rockets

In this section we will examine solid-propellant, liquid-propellant, hybrid-propellant, and multistage rockets. There are three main types of chemical propellant rocket. There are those that use solid fuels combined with a solid oxidizer in a single matrix, the so-called solid rockets. Then there are those that use separate liquid fuels and oxidizers, the so-called liquid rockets. And there are those that use a solid fuel with a liquid oxidizer (or vice versa), a design that gives a solid fuel rocket some of the benefits of a liquid rocket. These are the hybrid rockets. All rockets require onboard supplies of both a fuel and an oxidizer.

Fig. 4.3 The Soviet *Buran* space shuttle, together with its Energiya launch vehicle. Unlike the American Space Shuttle, Energiya used liquid-propellant strap-on boosters rather than solid rocket boosters (courtesy http://www.buran-energia.com)

Solid propellant rockets use a single rubbery compound bonded to the interior of the rocket casing itself. The inside of this solid "grain" is hollowed out in a specific shape to provide a chamber for combustion to take place at a certain rate and in a certain way. The grain contains both the fuel and the oxidizer mixed together, so that combustion can be sustained for as long as the propellant lasts. Once a solid rocket is ignited, there is no shutting it off; there are no valves and there is no other method of squelching the high energy combustion. The inside of the grain, along its entire length, serves as the combustion chamber, so that the propellant canister and the combustion chamber are one and the same.

One of the advantages of a solid-propellant rocket is that it can be stored for very long periods of time before it is used. Another advantage is the very simple design, which makes it easy to produce, possible to refurbish with a new rubber matrix, and very reliable. These items are so reliable, in fact, that the explosive bolts holding the Space Shuttle to the pad are blown just *before* the solid rocket boosters are ignited. Imagine what would happen if the solids did not ignite. In that case, the Shuttle would tip over and blow up, with no hope of rescuing the crew.

The major drawback to solid rockets is the fact that they cannot be turned off, throttled down, or powered back up. Although they can be refurbished, which involves repacking the case with new propellant, the cost and complexity of doing so are large compared to liquid refueling operations.

Liquid-propellant rockets (Fig. 4.3) use a variety of fuels and oxidizers, kept in isolated tanks until they reach the combustion chamber. Some of these are cryogenic, requiring extremely low temperatures and tank insulation – liquid hydrogen and oxygen are two examples. Some are hypergolic, igniting spontaneously on contact. Monomethyl hydrazine (MMH) and nitrogen tetroxide (N_2O_4) are examples of hypergolic propellants.

Some liquid propellants are very toxic; others are extremely corrosive. And some are ordinary fuels, such as gasoline or kerosene. The V-2 rocket used alcohol made from potatoes. The great advantage in liquid rockets is that they can be turned off or on, and throttled to provide a variable thrust. High specific impulse rockets often use cryogenic propellants during launch, while in-space reaction control system thrusters tend to rely on hypergolic propellants. The major disadvantage in liquid rocket engines is their complexity. The propellants have to be highly pressurized before being admitted to the thrust chamber, and valves have to work in extremes of temperature and pressure; nozzles have to be cooled. Above all, the engine must not malfunction during operation, because then the "controlled explosion" of the propellants could potentially become an uncontrolled conflagration. Helium tanks are sometimes used to pressurize and deliver propellants to the thrust chamber, avoiding the use of temperamental turbopumps. This increases the reliability of the liquid engine during space missions, but it also increases the weight of the vehicle.

A third class of rocket engine is the hybrid-propellant rocket. This is the type used by SpaceShipOne to win the Ansari X-prize. A hybrid rocket uses either a solid fuel and a liquid oxidizer, or vice versa. By using a liquid together with a solid, the engine can be throttled like a liquid-propellant rocket, but retains some of the simplicity and reliability of the solid rocket motor design. Like both solid and liquid rockets, hybrid engines can be dangerous if not handled carefully.

A multistage rocket is simply a series of smaller rockets stacked together. Only the tip requires an aerodynamic point, turning the lower stages into cylindrical sections. By using each stage in succession, and dropping off expended stages, the remaining stack of rockets can be accelerated to much greater velocities. The staging concept turns a regular rocket into a modular missile, with a greatly increased effective mass ratio, greater range, and higher speed. Orbital spaceflight is now within reach. As we will see in the next section, the staging concept is not limited to launch vehicles alone, but is used in modular spaceflight operations as well.

Modular Spacecraft

Unlike the history of aircraft development, the development of the rocket – from the simple artillery rockets of past centuries to the complex vehicles of today – has been characterized by several factors: (1) unpiloted designs, (2) throwaway pieces, and (3) modularized components. The relationship of these factors to one another is obvious, since an unpiloted projectile lobbed at the enemy would of course be thrown away, especially if it blew up at its destination. Modularized components also lent themselves to being discarded, especially if they improved the overall performance of the projectile. This process began with the addition of extra stages, as tubular missile sections were stacked one atop another. The technique was extremely successful in boosting a practical payload into low Earth orbit, especially if that payload was a satellite that had to be boosted only once. It therefore made sense to throw away the booster stages as soon as they had fulfilled their purpose.

This modularized approach was quickly extended to the spacecraft itself, when engineers realized that it was possible to discard certain components at certain points in the mission, to achieve overall success. Thus the retro-rocket pack affixed to the blunt end of the Mercury spacecraft was jettisoned just after burning its propellant, and just before atmospheric entry. The Gemini and Apollo spacecraft each had a crew cabin or Command Module (CM) and an unmanned section or Service Module (SM). The Russian Soyuz spacecraft has three sections: an Orbital Module, a bell-shaped Descent Module, and an unmanned Instrumentation and Service Module.

The Apollo Moon missions between 1968 and 1972 depended on modularized components. As each stage or module completed its assigned mission, it was left behind or cast off, and the crew carried on in the remaining spacecraft. A typical Lunar landing mission started with eight spacecraft pieces stacked and packed into the Saturn V, including the launch escape tower, and came back with only one, the conical CM. The Apollo-Soyuz Test Project in 1975 used a Docking Module, which allowed the Apollo Command and Service Modules to dock with the triply modularized Soyuz spacecraft (Fig. 4.4). Of the six modules used during the historic link-up, only two returned to Earth with their crews following the mission.

Most space stations (Fig. 4.5) consist of a large number of carefully designed modules, which plug into the station at various points. To build large structures in space, it has been necessary to build them piece by piece on Earth, and then loft the individual components into orbit one module at a time. The single exception to this rule was Skylab, a large three-man orbital workshop launched in 1973. It was not built up out of small modules at all, but was constructed from a converted S-IVB rocket stage. Skylab was probably the single largest piece of space hardware ever launched. It was, however, one huge module, boosted into orbit by the powerful Saturn V launch vehicle (Fig. 4.6).

The convenience and popularity of the modular technique in spaceflight can be explained by several factors. These include the all-important mass ratio, payload capacity, governmental largess, and practical design considerations. Let us take a

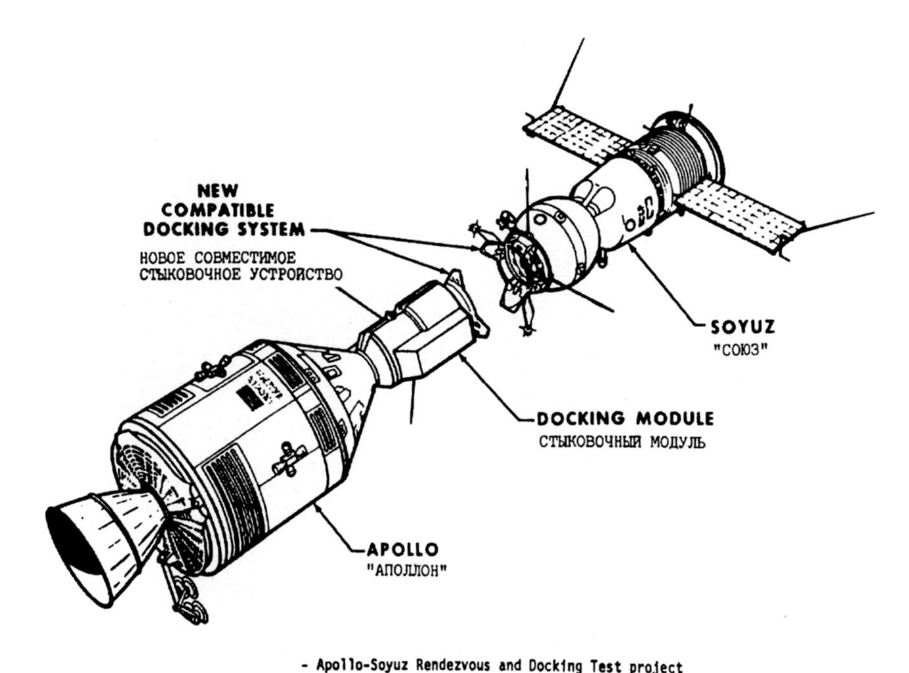

Fig. 4.4 Technical drawing of the Apollo-Soyuz Test Project vehicles, clearly showing the modularized components used during the historic American–Russian link-up in July 1975 (courtesy NASA)

quick look at the old Apollo Lunar mission architecture, which incidentally will be virtually replayed in a slightly modified, beefed-up, and updated version when NASA returns to the Moon around 2020.

The first three stages of the Saturn V launch vehicle were used to insert the spacecraft into low Earth orbit. There were five F-1 engines in the first stage burning kerosene and liquid oxygen, followed by five J-2 engines in the second stage and a single J-2 engine in the third stage, each burning liquid hydrogen and LOX. Upon low Earth orbital insertion, the third stage was not jettisoned, because it still had enough propellant remaining to insert the spacecraft into a trans-Lunar trajectory. By now the launch escape system (Fig. 4.8) had long been jettisoned. At the proper time and after appropriate checks of all systems, the third stage engine was restarted and boosted the remaining stack toward the Moon. At this point, the transposition and docking maneuver took place, in which the Command and Service Modules (CSM) separated from the third stage docking adapter, turned around, docked with the stowed Lunar Module (LM), and extracted it from its berth (Fig. 4.7). All this happened while the assembly coasted at 25,000 mph away from Earth. Ridding itself of the empty third stage, the Moon-bound spacecraft now consisted of four modules: the conical CM housing the three-man crew, the

Fig. 4.5 "Technical rendition" of STS-71 Shuttle orbiter docked to the Russian Mir space station. Note the contrast between the modular station and the nonmodular Shuttle (courtesy NASA)

unmanned cylindrical SM, the two-man LM ascent stage, and the unmanned LM descent stage with its four gangly landing legs. Each of these modules, or stages, had its own rocket engine, with the single exception of the CM. The CM had small thrusters only. Upon reaching the Moon, the high area-ratio Service Propulsion System engine was fired to decelerate the spacecraft into Lunar orbit. This was followed by undocking of the LM from the CSM, with two astronauts at the controls. Descent was made to the Lunar surface using the lone engine in the LM's descent stage, which would later serve as a launchpad for the ascent stage (Fig. 4.9). After

Fig. 4.6 A fictional example of a nonmodular space station, from the movie *2001: A Space Odyssey*. The station is apparently under conventional construction, rather than relying on modular architecture (courtesy NASA)

Lunar explorations were complete, the ascent stage engine was fired, leaving the descent stage behind. The two Moon explorers now rendezvoused with the orbiting CSM, and they transferred themselves and their Moon rocks into the CM. At this point the ascent stage was cast off, and the combined CSM returned to Earth, again using the Service Propulsion System engine. Just before impacting the Earth's atmosphere at more than 36,000 ft/s, the cylindrical SM was set adrift, so that the conical CM with its three-man crew and lunar samples could make a safe and free return to Earth.

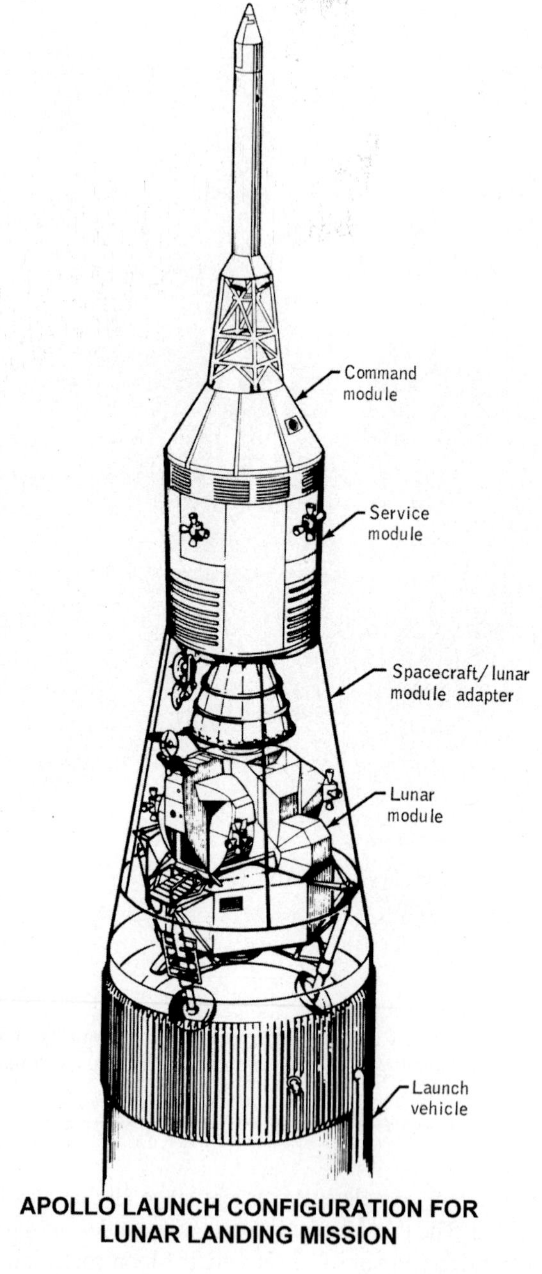

Fig. 4.7 Cutaway drawing showing the modular components of the Apollo spacecraft and stowed Lunar Module (courtesy NASA)

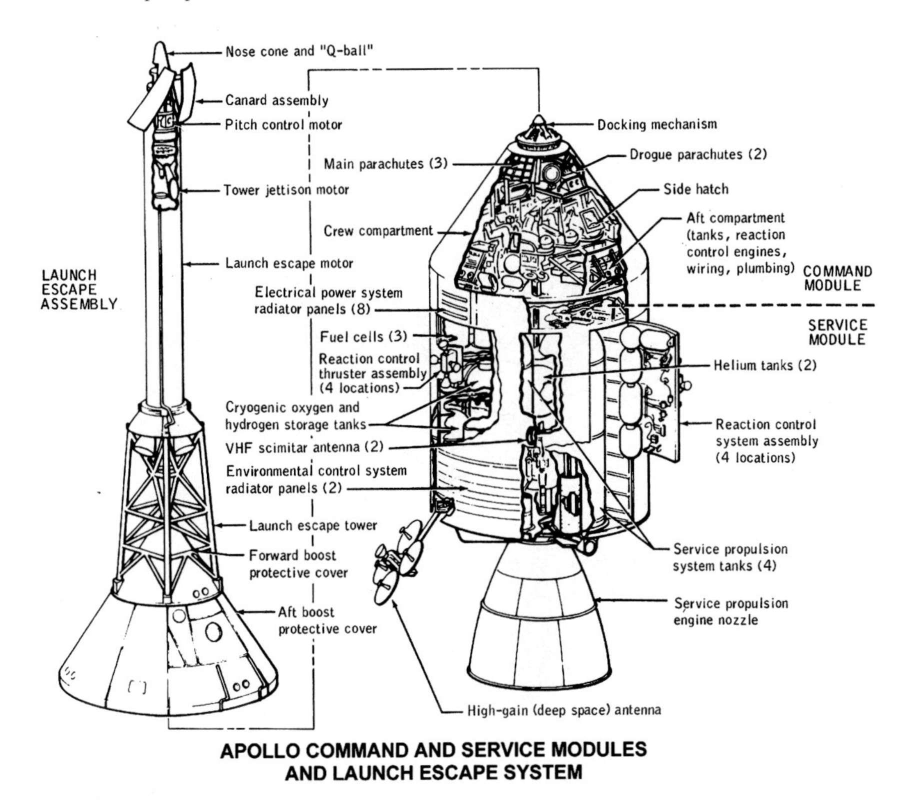

Fig. 4.8 Technical drawing of the Apollo Command and Service Modules, together with the launch escape system, which was jettisoned prior to reaching orbit (courtesy NASA)

What is the relationship of missiles to modules? Launch vehicles derived from ballistic missiles depend on modular construction to work at all. Each stage is, in effect, a module. Likewise, each module is, in fact, a stage. This design methodology makes spaceflight much easier than would otherwise be the case, but it also boosts operational costs to the point that a vicious circle results. These costs – and the *expectation* of continued high costs – dictate the planning of many fewer missions while demanding much higher reliability. This drives up the complexity and cost of all components, which in turn leads to low launch frequency, thereby continuing this malicious cycle.[3]

Rockets vs. Spaceplanes

The best thing going for rockets today is the fact that we have them, we understand them, and we know how to fly them with reasonable proficiency. With modular designs and stacked stages, it is a fairly straightforward procedure to attain the energies

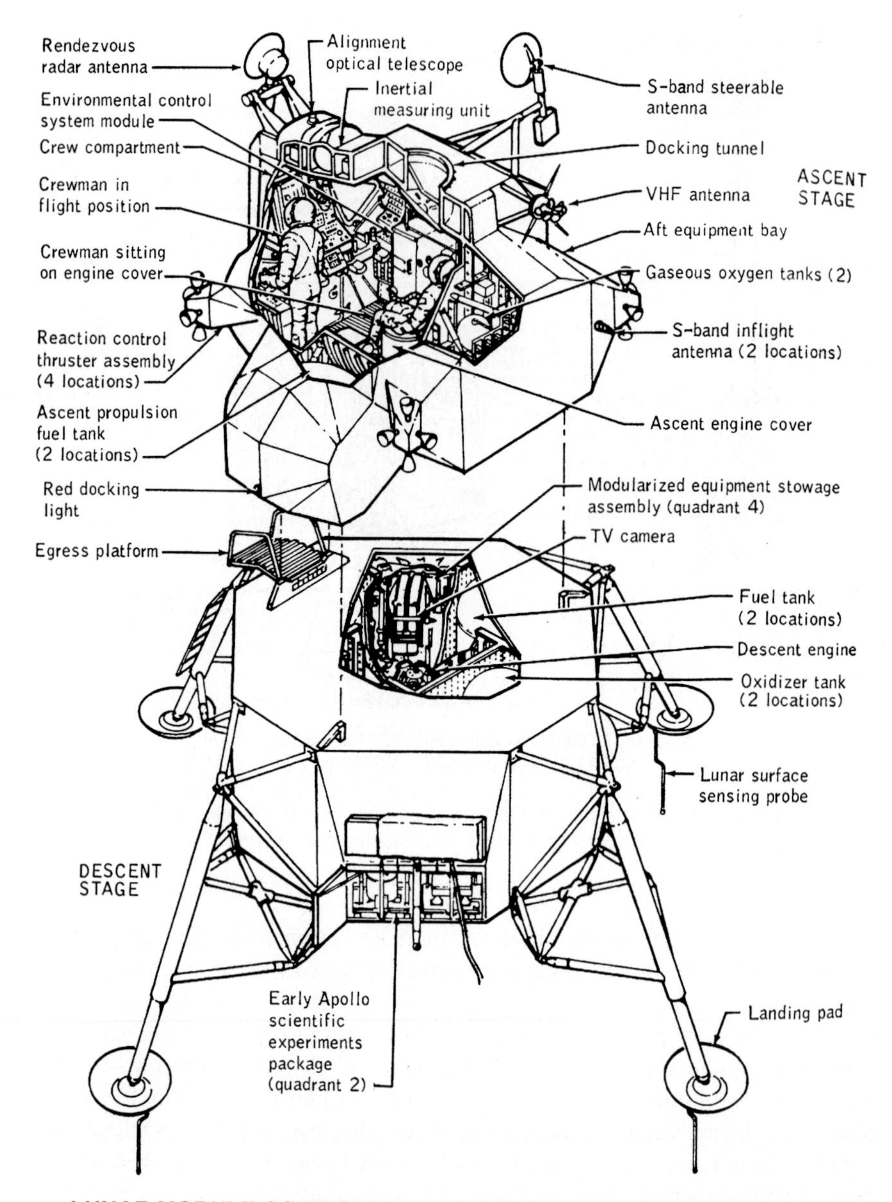

Fig. 4.9 Detailed drawing of the two-piece Apollo Lunar landing vehicle, consisting of descent and ascent modules, each with its own rocket engine (courtesy NASA)

required for spaceflight. Conventional launch vehicles enjoy simple construction in which most flight loads are longitudinal. From a purely engineering standpoint, the designs are simple, robust, and lightweight. They work.

The major factor arguing against the ballistic booster is the waste and the inherent cost associated with every launch. Not only is the launch vehicle, along with its complex rocket engines, sacrificed on every launch, but Earth's atmosphere is ignored. It is not used for its free lift, its propulsive potential, or its oxygen content. It is seen only as a barrier to be crossed. Rockets throw away their gas tanks and engines every time they are launched. Can you imagine doing that with the family car?

The Achilles' heel of all conventional vertical launch vehicles is their one-time usability. This single factor leads immediately to extremely high costs and lower reliability. The space launch industry is typified by a 2% failure rate in its launch vehicles. Such failure rates in the airline industry would never be tolerated. Millions of passengers fly every day, and only occasionally does an airliner crash. With a 2% failure rate, a private pilot with 300 h total time would be dead six times over. The low reliability of ballistic launch vehicles is due to their high cost and consequent inability to be tested like reusable aircraft.

Another problem with ballistic launch vehicles is the market. The only customer for vertical launchers has been the occasional satellite, interplanetary probe, or manned mission. Space tourists have been very few, and so far have had no impact on flight frequency or improved launch architecture. All of this is about to change, with the advent of the spaceplane.

Spaceplanes will be far safer than vertical launch vehicles, precisely because of their reusability. Their wings and wheels (Fig. 4.10) will allow a development and flight test program to proceed the same as for any other aircraft. Thousands of hours will be spent putting the vehicle through its paces, gradually expanding the flight envelope, and identifying inherent design problems that can then be corrected. The greatest challenge will be in upgrading the performance and reliability of the rocket engines, but even this task will be attainable, in principle. Reusability and the resulting increased flight frequencies are fundamental in developing good, reliable, relatively inexpensive space vehicles.[4]

Spaceplanes will someday completely replace missiles and modules. It may take a long time before this vision is completely realized. For the next several decades, rockets will continue to be the mainstay of space access, especially where heavy lift is required. But for the short term, spaceplanes will blaze the way for ordinary people to enter space as tourists. The first space tourists will pay in the neighborhood of $200,000 each for the privilege of experiencing a few minutes of weightlessness at altitudes above 100 km. These brave souls will be the first private suborbital spacefarers.

We already have suborbital spaceplanes. SpaceShipOne flew in 2004, twice within 14 days, winning Scaled Composites the $10 million Ansari X-prize. SpaceShipTwo, an enlarged version of SS1, is being built even now under a veil of extreme secrecy, and will soon make its first test flights. Other companies are working on spaceplanes of their own. Some companies envision using carrier aircraft to launch spaceplanes from altitude. Others plan a simple runway takeoff without the

Fig. 4.10 Wings and wheels are the key to repeated flight test of new spaceplanes. This is the HL-20 lifting body parked on the ramp at Langley Research Center in 1992 (courtesy NASA)

expense of operating a separate large mothership. Some companies like hybrid rocket engines; others prefer only liquid propellants. Some companies are building their own engines, much as the Wright brothers did over a century ago. When will these suborbital vehicles appear in large numbers? They could appear any day now. By the time you read these words, they could already be flying on a regular basis.

The greatest strength of the spaceplane is the fact that it has wings and wheels, which of course translates immediately to reusability. Reusability leads to more flights, better testing, lower costs, and greater reliability. Reliability, in turn, leads to more confidence and greater flight frequency, which in turn leads to more flight experience, more design improvements, better vehicles, and even more flights at lesser cost. This self-perpetuating circle is just the opposite of the vicious circle of rocketry described earlier, in which an expectation of higher costs leads to fewer flights with a demand for greater reliability, which in turn leads to higher costs and even fewer flights. Striving for improved reliability without increased flight frequency is clearly not the way ahead.

Superior aerospace vehicles, as spaceplanes will be, demand superior design, better materials, and optimal flight profiles. Apparent weaknesses will be turned into strengths, turning for example a vehicle whose propellant tanks are too large into a space tanker.

In traditional rocket science, dead weight is regarded the same way one might regard a 100-lb tumor clinging, leachlike, to one's body. Yet we have already seen that the wings and wheels of a mature spaceplane are critical in ensuring that the most vital aspect of the spaceplane – *its reusability* – is preserved. So from the perspective of the spaceplane, wings and wheels are not dead weight at all, but essential parts of the ship. And this should be true regardless of how far into the ocean of space that ship sails.

Spaceplanes to the Moon

The fully mature spaceplane of the future will have the ability to take off from any spaceport on Earth, cruise into low orbit, refuel, and fly its passengers and cargo to the Moon. Its cargo may comprise not only bulk dry goods and supplies for a Moon base, but also precious rocket fuel for the base's day-to-day operations. The mature spaceplane would best be utilized in this way, serving as space tanker both initially *to* and eventually *from* the Moon.

In his book *Return to the Moon*, Apollo 17 geologist-astronaut Harrison H. Schmitt reveals that production of helium-3 from Lunar soil will release a host of other elements and compounds, including helium-4, nitrogen, carbon monoxide, methane, "large amounts of hydrogen," and water. Helium-3, which is extremely rare on Earth but is common on the Moon because of the Solar wind, could be used in fusion plants to power the energy demands of the twenty-first century.[5] But consider the list of by-products: "Large amounts of hydrogen," the best rocket fuel in the Universe; water, which can either be consumed or hydrolyzed into hydrogen and oxygen – the best rocket propellant combination in the Universe; methane, another rocket fuel or source of energy, in any case; and nitrogen, an inert element that can be combined with oxygen to provide breathable air.

Now the pieces of the puzzle are beginning to fall into place. The Moonplane, with its enormous propellant tanks and modest cabin, can be filled with Lunar rocket propellant produced as by-products of helium-3 production. Transporting these precious fluids from the Moon to low Earth orbit requires only a small fraction of the energy

Fig. 4.11 The tail-less X-36 parked on the ramp at Dryden Flight Research Center, California, in 1997. Will future spacecraft bound for the Moon look like this? (courtesy NASA)

that it would take to transport the same cargo in the opposite direction, because Earth lies at the bottom of its own gravity well. Gravitationally speaking, it is "downhill" from the Moon. Moreover, the spaceplane is ideally suited to make this downhill run, because it has the unique ability to decelerate in Earth's atmosphere, emerge back into the vacuum of space, and settle into a low Earth parking orbit. Propellants are delivered to the orbital depot, and the spaceplane reenters the atmosphere – this time in earnest – and returns to its home spaceport.

Spaceplanes to Mars?

If Lunar spaceplanes become a reality in the future, then what would prevent there from flying farther afield? What about Mars? After all, Mars has an atmosphere, of sorts, that spaceplanes could make some use of.

Indeed. The very same arguments that favor spaceplanes flying to the Moon can be made in support of spaceplanes flying to Mars. These involve chiefly the space tanker potential they bring with them, as well as their winged versatility. Some sort of lifting body or winged structure vastly increases the versatility of all spaceplanes, compared to other designs. Spaceplanes can enter planetary atmospheres and decelerate with far less stress than their blunt-body cousins, the capsules. This means that spaceplanes will be much more comfortable to their passengers than any other kind of spacecraft, especially during atmospheric entry.

Granted, the logistics of manned Mars flight by spaceplane, as with any spacecraft, are very complex. In the final analysis, the successful Mars mission will probably involve sending unmanned cargo craft to the planet well ahead of the crew, who will then follow in a small, light, fast spacecraft. Again, the availability of proven and versatile spaceplanes may well turn out to be just what is needed to send crews to the Red Planet.

Imagine this scenario. The year is 2040. Lunar spaceplanes have been perfected, and two are now being outfitted for the first manned Mars mission. They are parked in Lunar orbit, and the provisioning is almost complete. Several automated cargo vessels have already arrived at Mars and are standing by at the staging area at the northwest rim of Valles Marineris. The Mars-bound spaceplanes have been specially modified with dorsal docking collars, so that a special extendable truss can be deployed between the two spaceplanes during the transfer orbit. This will allow the generation of artificial Mars gravity onboard both spaceplanes by mutual rotation around their center of gravity, and allow crews to pay visits to their sister craft during the journey. Each Marsplane is fully loaded with propellants in anticipation of a relatively quick 2½-month crossing.

When the Marsplanes reach the planet, they will separate, stow their collapsible tunnels, and aerobrake through the Martian atmosphere. Possibly making several skip-passes to gradually bleed off airspeed, the craft will glide to landings at the staging area and set down gently using their specially fitted belly engines. Automatic cameras at the base record the arrival. First one, then a second interplanetary bird comes swooping out of the red sky. They make a high-altitude pass over the base, verify their position, and ignite their ventral thrusters. A last-second pitch up of the nose on final approach, they flare to landing and make a smooth touchdown in the Martian dust. Within minutes, the second bird makes a repeat performance.

This could actually be the way we go to Mars – by spaceplane rather than by supercapsule. Before such a reality can materialize, it has to crystallize in the imagination.

References

1. Frederick I. Ordway et al., *Basic Astronautics*. Prentice-Hall, Englewood Cliffs, NJ, 1962, p. 13.
2. Willy Ley, *Missiles, Moonprobes, and Megaparsecs*. Signet, 1964, Chap. 1.
3. James R. Wertz and Wiley J. Larson, eds., *Reducing Space Mission Cost*. Kluwer, Boston, 1996, p. 56.
4. David Ashford, *Spaceflight Revolution*. Imperial College Press, 2002.
5. Harrison H. Schmitt, *Return to the Moon*. Praxis, 2006, p. 79.

Chapter 5
Crawling into Suborbit: The Baby Spaceplane

For those who would defame the development of spaceplanes, especially suborbital ones, it behooves me to insert a few words here in their defense. It may seem that such vehicles are nothing more than the space toys of rich entrepreneurs, a contrivance of those with too much time and too much money on their hands. In truth, suborbital spaceplanes represent the first toddling steps in the development of real spaceships. We are talking about vessels that, marshalling immense energies, will one day ply the interplanetary spaceways and sail the cosmic sea. We are not talking about modified ballistic missiles used to start spacecraft coasting toward their eventual targets. Suborbital spaceplanes really are baby spaceships, and as babies their abilities are understandably limited. It is entirely fitting that they would be helped into flight by the strong wing of a mothership.

And yet, the potential of the infant spaceplane is far greater than even the most powerful aircraft ever conceived. For airplanes, no matter how large, no matter how fast, no matter how far they may fly, can never leave Earth. It may well be that when the history of our era is written in the centuries to come, the ballistic method of entering space will be seen as a temporary aberration on the true path to the development of a sustainable space-based civilization. Only spaceplanes, with their potential to leave Earth as often as airplanes today leave the ground, can make a real difference in expanding our civilization into the far reaches of space.

You already know what makes a spaceplane – a plane that can fly in space. But what defines a suborbit? Just as a submarine is a boat that floats under water, a suborbit is a trajectory below an orbit, specifically one with a speed below that of orbital velocity. Suborbital space vehicles are either too low, too slow, or both. If they are too low, then they are still inside the sensible atmosphere, and aerodynamic drag will soon bring them down. If they are too slow, then gravity will play its part in dragging the spacecraft back to Earth, even in the absence of an atmosphere. Typical suborbits reach an apex, or apogee, outside the atmosphere, but with insufficient speed the orbit quickly succumbs to the will of gravity and arcs back toward the ground. Nevertheless, passengers and pilots of suborbital craft can still experience a few minutes of real spaceflight conditions. These include the weightlessness of suborbital free fall, the inkiness of a daytime sky, the curvature of Earth, and the silence of space. In the very near future, thousands of space tourists will experience the joys of suborbital spaceflight.

M.A. Bentley, *Spaceplanes: From Airport to Spaceport*,
doi: 10.1007/978-0-387-76510-5_5,

The Mercury Redstone

The first two manned space missions launched by the United States were suborbital lobs. On 5 May 1961, NASA astronaut Alan B. Shepard rode a Redstone rocket (Fig. 5.1) inside his tiny Mercury spacecraft from the coast of Florida out into the Atlantic Ocean, to a distance of some 302 miles. The flight lasted only 15 min 22 s.

Let us now take a look at exactly what Shepard experienced and compare it to some other suborbital flights. Although the Redstone carried insufficient propellant to reach an orbital speed of 5 miles/s, its engines had enough thrust – some 78,000 lb – to accelerate the Mercury spacecraft to a speed of 5,180 mph, or 1½ miles per second. This was just under one third of the velocity needed to maintain orbit. As the rocket lost more and more mass with the burning of its propellant, the vehicle

Fig. 5.1 Launch of the suborbital Mercury Redstone rocket with Alan B. Shepard, the first American in space, May 5, 1961. This flight took place less than a month after the Russian cosmonaut Yuri Gagarin became the first man in space on April 12, 1961 (courtesy NASA)

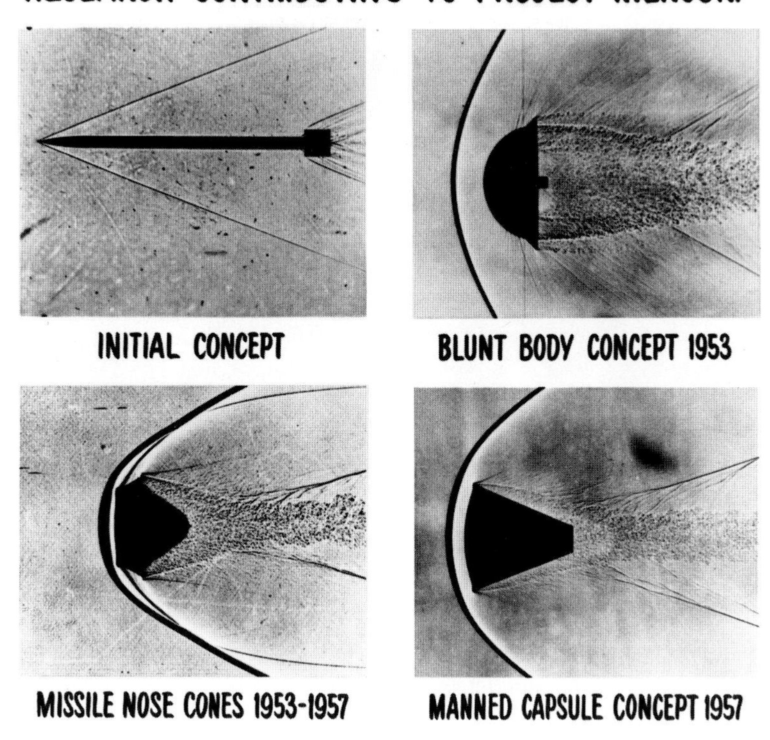

Fig. 5.2 Shadowgraph research images of various reentry vehicle shapes, illustrating the bow shock waves experienced by returning spacecraft (courtesy NASA)

began to accelerate more and more rapidly. This caused a gradual build-up of G forces. Starting at the normal 1 *G* just before liftoff, Commander Shepard experienced 6 *G*'s, or six times the pull of gravity, before the engines cut off. He and his spacecraft then experienced about 5 min of weightlessness as they arced through apogee at 116½ miles altitude. This is well above 100 km, the official border of space. The fun was just about to begin, though, for when the bell-shaped spacecraft turned around and reentered the atmosphere (Fig. 5.2), its low lift-to-drag ratio and greatly reduced mass (the launch vehicle was no longer attached) meant that it would slow down much more rapidly than the Redstone had accelerated it initially. With the blunt end pointed forward, the astronaut experienced a crushing 11 *G*'s of deceleration. He was therefore pushed into his seat both on the way up and on the way down. Shepard's capsule made the final descent by parachute and splashed down in the Atlantic, to await recovery by helicopter (Fig. 5.3).

On July 21 of the same year, Gus Grissom made a similar flight, reaching an altitude of 118.3 miles, experiencing 5 min 18 s of weightlessness, and 11.1 *G*'s on reentry. These were the first Americans to enter space, and the only human beings,

Fig. 5.3 America's first spaceman, the suborbital Alan Shepard, being recovered by helicopter from the Atlantic Ocean. He later went on to become the only Mercury astronaut to walk on the Moon, as Commander of *Apollo 14* in February 1971 (courtesy NASA)

to date, to be launched on suborbital trajectories by ballistic missiles. The two Russian cosmonauts who entered space that year – Yuri Gagarin on April 12 and Gherman Titov on August 6 – were boosted by bigger rockets that threw them all the way into elliptical orbits about our planet. Has anyone else been to space in a suborbital vehicle? Yes, they have, in a baby spaceplane called the X-15.

The North American X-15

Suborbital winged spaceflight began in 1962 with the X-15 rocket research aircraft, which had already been flying below space altitude since 1959. Ballistic missiles had lobbed four men into space the previous year, and had done so with greater energy than the X-15 could muster. And yet the X-15 could be flown over and over, because it had wings. But it had something most aircraft did not have, and that was a rocket engine, enabling it to operate outside the sensible atmosphere.

The X-15 research aircraft routinely entered regions of the upper atmosphere where its wings were aerodynamically ineffective. It therefore had to be equipped with small thrusters to maintain control during the highest parts of the flight. These were located on the wingtips for roll control and in the nose for pitch and yaw control.

The US Air Force awarded astronaut wings to X-15 pilots who flew above 50 miles. NASA, on the other hand, required its pilots to fly above 62 miles – 100 km – before considering them real astronauts. The X-15 program included pilots from prime contractor North American Aviation, the US Air Force, the US Navy,

Fig. 5.4 The X-15 just after landing on the dry lake at Edwards AFB, with its B-52 mothership flying over (courtesy NASA)

Fig. 5.5 Space pilot Neil Armstrong with the X-15, about 1960 (courtesy NASA)

and NASA. As a result, some X-15 pilots earned astronaut wings, while others did not. One X-15 pilot who earned astronaut wings was Air Force Capt. Joe H. Engle, who later went on to pilot the Space Shuttle for NASA. One who did not was civilian pilot Neil Armstrong (Fig. 5.5), who only reached 39 miles. Ironically, he later became the first human to walk on the Moon, in July 1969.

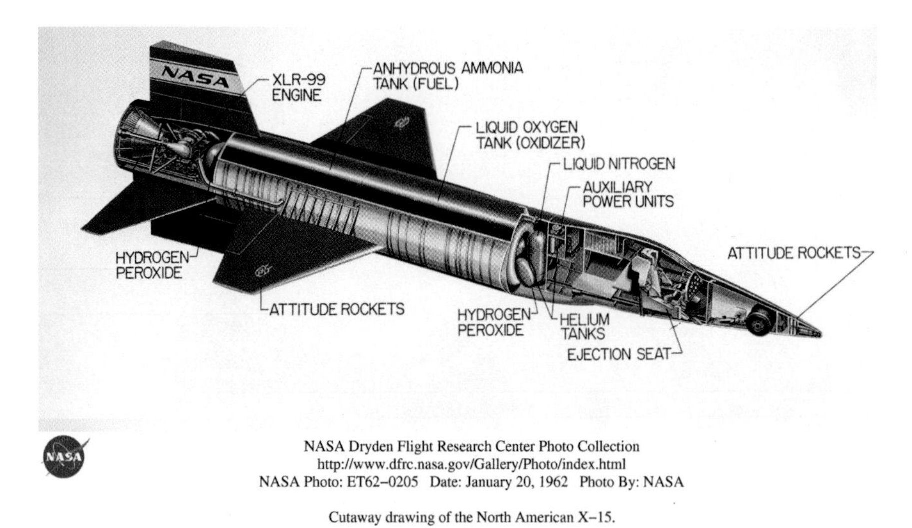

Fig. 5.6 Cutaway drawing of the X-15 showing internal structure and general arrangement of propellant tanks and engines, and landing gear. The rear landing gear was a set of retractable skids (courtesy NASA)

The X-15 was the ultimate rocket research aircraft of all time. Like its predecessors, it was launched from a large carrier aircraft (a modified B-52) and was flown by a single pilot. Its engine, the XLR-99, provided a thrust of 57,000 lb. It had very stubby wings, and tested the limits of flight at the edge of space. Three X-15s were built (Fig. 5.6) by North American Aviation, making a total of 199 flights between 1959 and 1968. During the research program, there were three crash landings, but only one fatality. One craft broke in half on landing, but its pilot, Scott Crossfield, survived. Another blew up on the test stand with the same pilot, but again he survived. A third craft suffered loss of control and made a high-speed landing in which the vehicle flipped over (see Fig. 5.7). The pilot, Jack McKay, was trapped inside the wreckage for 4 h after banging his helmet on the runway. Amazingly, both craft and pilot flew again, many times. The only pilot to lose his life was testing a procedure in which a computer was allowed to fly the plane. After successfully recovering from an extreme altitude 3,000 mph spin, a severe pitch oscillation ensued, knocking out USAF Captain Mike Adams because of excessive G forces, and the vehicle disintegrated before he woke up. He had just exceeded the 50-mile space altitude, as recognized by the US Air Force, and was posthumously awarded astronaut wings.

To really appreciate the success of the X-15, it is useful to compare some of its statistics with that of the Space Shuttle. By the time the Shuttle had made 115 flights, two vehicles had been lost, with their crews of seven each. The X-15 made 199 flights with the loss of one vehicle and one man. The shuttle has averaged six

Fig. 5.7 On November 9, 1962, X-15 pilot Jack McKay experienced multiple vehicle failures, forcing a crash landing. After the engine became stuck at 35%, the planned flight test was aborted. Landing flaps then failed to extend, resulting in a higher than normal-speed approach, and the left skid failed on touchdown, causing the vehicle to tumble. Yet because this vehicle had wings and direct human control, both plane and pilot flew again (courtesy NASA)

flights per year, while X-15 averaged 22. The X-15 provided much of the data subsequently used in NASA's humans-in-space program.

The X-15's highest altitude flights had something in common with the first two manned American spaceflights. In both cases, flights were suborbital. How does the X-15 compare to these Mercury Redstone flights? Table 5.1 presents a few data points, which include NASA's first two manned spaceflights, all X-15 flights above 50 miles, and the five fastest flights of the X-15.

There are two matters of significance in these data. The first is that there were only two flights of a manned Redstone rocket – ever. The Redstone rocket had been developed by the US Army as a ballistic missile in the Redstone arsenal. It was modified for use by Project Mercury, because NASA knew that it could boost a small spacecraft into suborbital space. The Redstone had also been used to fly Ham, the space-chimp (Fig. 5.8), on a flight similar to Commander Shepard's, on 31 January 1961. Including this flight, three separate launch vehicles and three separate spacecraft were used for a total of three suborbital flights. By contrast, the winged X-15 made 199 flights in 9 years, using just three vehicles, including 15 flights in 1961.

Winged spacecraft are reusable because of their wings. Extensive testing of any flight vehicle requires that vehicle to have wings so that it can be reused and tested over and over again. This is the fundamental difference between ballistic capsules and spaceplanes. And this is why spaceplanes will be inherently safer than missiles and modules.

Fig. 5.8 Ham, the astro-chimp, was the first chimpanzee to enter space, blazing the trail for NASA's human astronauts. Meanwhile, the Russians were sending dogs, like Laika, into space (courtesy NASA)

Table 5.1 Comparison of Mercury Redstone and X-15 suborbital flights[1,2]

Date	Flight	Max. altitude (miles)	Max. speed (mph)	Pilot
05 May 1961	MR-3	116.5	5,180	Alan Shepard
21 July 1961	MR-4	118.3	5,200[a]	Gus Grissom
09 Nov. 1961	X15-45	19.24	4,093	Robert White
27 June 1962	X15-59	23.43	4,104	Joseph Walker
17 July 1962	X15-62	59.61	3,832	Robert White
17 Jan. 1963	X15-77	51.46	3,677	Joseph Walker
27 June 1963	X15-87	53.98	3,425	Robert Rushworth
19 July 1963	X15-90	65.87	3,710	Joseph Walker
22 Aug. 1963	X15-91	67.08	3,794	Joseph Walker
05 Dec. 1963	X15-97	19.13	4,018	Robert Rushworth
29 June 1965	X15-138	53.14	3,432	Joe Engle
10 Aug. 1965	X15-143	51.33	3,550	Joe Engle
28 Sep. 1965	X15-150	55.98	3,732	Jack McKay
14 Oct. 1965	X15-153	50.47	3,554	Joe Engle
01 Nov. 1966	X15-174	58.13	3,750	William Dana
18 Nov. 1966	X15-175	18.73	4,250	Pete Knight
03 Oct. 1967	X15-188	19.34	4,520	Pete Knight
17 Oct. 1967	X15-190	53.13	3,869	Pete Knight
15 Nov. 1967	X15-191	50.38	3,570	Mike Adams
21 Aug. 1968	X15-197	50.66	3,443	William Dana

[a]Estimated

Although the Redstone lifted off from sea level, it easily – and admittedly – outperformed the X-15 in both speed and altitude. Ballistic missiles are inherently good at lobbing fast projectiles. While the X-15 and SpaceShipOne each had zero forward speed at the apogees of their respective suborbital trajectories, Mercury had a significant suborbital velocity. This is the reason it landed 300 miles downrange from the launch point, and a large part of the reason Shepard subsequently experienced such high G forces.

The fastest X-15 flight was number 188 on 3 October 1967, which reached 4,520 mph. This was only 660 mph slower than Alan Shepard's suborbital lob 6 years earlier. This X-15 flight carried extra propellant tanks and had the longest burn time of its rocket engine, more than 140 s. The fastest X-15 flights did not reach the highest altitudes, however. Pete Knight's unofficial X-15 speed record of 4,520 mph never took him or his craft above 20 miles, compared with Alan Shepard's apogee of 116½ miles. Again, ballistic missiles outperform aircraft in both speed and altitude, but not in the vital areas of reusability and reliability.

Although the X-15's performance cannot match that of even the relatively underpowered Redstone rocket, the fact that it had wings meant that it could be flown time and time again. And this is the true value in all winged space vehicles. This point will be stressed repeatedly throughout this book. Reusability in flight requires wings. The reason spaceplanes have wings is so they can fly, yes, but more important, it is so they can fly *again*. This, in turn, leads to much greater reliability and safety because of the additional flight experience gained. In the case of the X-15, the engineering data gleaned from 9 years' worth of flight experience proved invaluable in the design of the Space Shuttle in the following decade. This research program could not have been carried out without a reusable, reliable vehicle such as the X-15.

What steps need to happen to bridge the gap between vehicles such as the X-15 and future orbital spaceplanes? There is more than one answer to this question, and it depends on our vision of the future. The process of gradual improvement in spaceplanes is already underway. As spaceplanes are enlarged, they will have the capability of transporting more passengers to greater speeds and altitudes. Yet, the road to space is long and arduous. The energy required to reach orbit is 30 times greater than what is required to reach space height, because orbital energy is mainly in the form of speed.

For a spacecraft in a circular orbit 200 miles, or 1 million ft, above Earth, the specific energy due to altitude is gravity times height, gh, or roughly 32,000,000 ft²/s². When an actual orbital velocity of 25,000 ft/s is added into the equation, the specific energy ($gh + \frac{1}{2}v^2$) becomes roughly 344,000,000 ft²/s². Fully 90% of this is in the kinetic or "velocity" term – that little $\frac{1}{2}v^2$. SpaceShipOne reached an altitude of 112 km on 4 October 2004, but by the time it reached this altitude, it had lost all of its speed, just like the X-15. Its specific energy at burnout can easily be found by calculating its specific potential energy at apogee, because this is where the speed and the kinetic energy went to zero. This is simply gh or $(32\ \text{ft/s}^2) \times [(112\ \text{km})/(0.0003048\ \text{km/ft})] = 11{,}700{,}000\ \text{ft}^2/\text{s}^2$. Dividing the first result by the second, we have

344,000,000/11,700,000, which is nearly 30 times as much energy to reach space height *and* orbital speed as required to reach space height only. This means spaceplanes need to be improved by a factor of 30 before they will reach low-Earth orbit. At present, spaceplanes are little more than manned sounding rockets.

How long will this process take? Let us take a look at the altitudes and speeds associated with the Wright *Flyer*. The original Wright *Flyers* had airspeeds in the 10–30-mph range. The X-1 broke the sound barrier – about 700 mph – in October 1947. This pivotal event required speeds roughly 30 times what the Wright brothers could achieve. So it took just under 44 years. An increase in speed is tantamount to an increase in energy, because as we have seen, for a spacecraft in a 200-mile orbit, 90% of a spacecraft's energy is tied up in speed. If the development of the spaceplane takes place at the same rate as that of the airplane, then the first orbital spaceplane should be flying by the year 2047.

Costs vs. Benefits

The developmental costs for suborbital spaceplanes are not insignificant. SpaceShipOne cost at least $20 million to develop from conception to flying object, winning back half that amount in the form of the X-Prize. The real benefit in terms of cost comes from repeated operations once spaceplanes enter regular service. If a single spaceplane can make 100 flights over a period of several years, the developmental costs can be amortized over that period, and costs per flight can be significantly reduced compared to one-shot missiles. Repeated operations, of course, depend on wings and wheels rather than staged rockets, modular spacecraft, and blunt reentry vehicles.

Space tourists will gladly pay large sums for the privilege of being flown into the space environment. They will effectively fund the development of better and better spaceplanes and ensure continual upgrades. In time, spaceplanes will gradually increase their capabilities and improve their safety records, until the advanced, single-stage-to-orbit spaceplane is finally developed. In this way, the huge costs to develop these vehicles will be shared by thousands of paying passengers, who will literally be funding their own future safe access to space in superior, reusable spaceships. These are the real benefits, but it seems it will require privately owned businesses rather than wasteful government programs, to make it happen.

Launch Methods

How does a suborbital spaceplane launch? There are two obvious methods: air launch and self-launch. Air launch is the method used in the testing of most high-speed rocket planes, including the early rocket-powered X-planes, the NASA lifting bodies, and the X-15. It was also the method used by SpaceShipOne. This technique allows the spaceplane to "light off" from 40,000 or 50,000 ft and some initial non-zero airspeed, giving it an advantage in terms of its flight energy. The total energy of a spaceship, you will recall, depends on the sum of potential and kinetic energies, which in turn depend on altitude and velocity. So the air-launched vehicle can achieve space height much easier than its self-launched counterpart.

On the other hand, if spaceplanes are to someday mature into the capable space transports of tomorrow, then those that launch themselves now may be at a distinct advantage over their overly coddled air-launched cousins. The child who has training wheels may ride his bicycle first, but the one who never has them may ultimately prove to be the better, faster rider. XCOR Aerospace plans to launch its Xerus spaceplane from a runway, under its own power.

Safety

Spaceflight is a dangerous business, in space, on the ground, and in between. One of the reasons for this is the nature of the volatile chemicals used to power the engines. That volatility comes about as a result of the high energy content of the propellants, which in turn is necessary because of the high energies required for spaceflight. Understanding the causes of every mishap and every near mishap in the history of rocketry and spacecraft engineering is vital in building the spaceplanes and spaceflight infrastructure of tomorrow. Indeed, entire volumes have been written on this topic for this very purpose.[3,4]

Safety, simplicity, and reliability are all interrelated. Simplicity of operations, further, influences the economics of the spaceplane. And only the most economical spaceplane will survive the market.[5] Taking its lead from the airline industry, the successful spaceplane must use horizontal take-off and landing and some kind of very reliable air-breathing engine. In this way it can operate independently of a carrier aircraft, and need not use rockets from the runway (Fig. 5.9). Noise concerns alone would otherwise conspire to ground the spaceplane before it could ever begin operations.

Fig. 5.9 Douglas Skyrocket takes off with small "jet assisted take-off" rockets in 1949 (courtesy NASA)

The Tourism Market

Suborbital spaceplanes are already being built, with the financial backing of Richard Branson and his Virgin Galactic Spacelines. The technical expertise is being supplied by Scaled Composites of Mojave, California, who has the experience in designing and building SpaceShipOne. Other companies are also working on the suborbital spaceplane concept, notably Bristol Spaceplanes and XCOR Aerospace. Although the Bristol Spaceplanes Ascender is planned to be the second stage in a two-stage-to-orbit design, the Ascender itself could easily serve as a suborbital spaceplane, offering tourist rides into deep space. XCOR Aerospace's Xerus spaceplane is designed to take off under its own power from a runway and fly into suborbit. The company has identified three markets: (1) space tourists, (2) microgravity experiments, and (3) small payloads to be boosted into orbit from the edge of the atmosphere. XCOR's methane-fueled rocket engines have already achieved a good operating and safety record. Of these three potential markets, it is clearly the prospect of space tourism that will provide the most paying customers. Small markets undoubtedly exist, as well, for upper atmospheric or microgravity experiments and small payloads requiring a boost into orbit. But in terms of sheer numbers, these will pale in comparison to the legions of adventurous souls who will gladly pay a small fortune for the experience of their lives. These trailblazing space tourists will understand perfectly well that mature spaceplanes will take time – and money – to develop. This development process will begin with the suborbital space lob: a few moments of exhilarating acceleration, a few minutes of weightlessness, and a peek at our planet from the inky realm of space.

Fig. 5.10 The Skylon spaceplane concept, as it may appear parked in its hangar (courtesy Reaction Engines Limited)

Spaceplanes are here to stay, but they have a long way to go before they mature into what could be called advanced spaceplanes (Fig. 5.10 and Chap. 9). They need to harness energies about 30 times greater than what either the X-15 or SpaceShipOne was capable of. Can it be done? It took 44 years for airplanes to advance from the 30-mph Wright *Flyer* to the 700-mph X-1. This represents a 30-fold increase in speed and energy, analogous to what is required of spaceplanes. There is much work to do, and plenty of challenges to keep engineers busy well into the future.

References

1. Reginald Turnill, *The Observer's Spaceflight Directory*. Frederick Warne, London, 1978.
2. Dennis R. Jenkins and T.R. Landis, *Hypersonic: The Story of the North American X-15*. Specialty Press, 2003.
3. David J. Shayler, *Disasters and Accidents in Manned Spaceflight*. Springer-Praxis, 2000.
4. David M. Harland and R.D. Lorenz, *Space Systems Failures*. Springer-Praxis, 2005.
5. David Ashford, *Spaceflight Revolution*. Imperial College Press, 2002, Chap. 13.

Chapter 6
Going Ballistic

As we saw in the last chapter, the easiest way to give a projectile enough energy to enter space is by using a ballistic missile. Although they are perhaps the most effective and certainly the quickest way to achieve spaceflight, ballistic missiles are also the most costly. Government-run space programs have always used ballistic missiles, because funding was seldom an issue. This was true especially during the early years of the Space Age. Furthermore, these programs have never been bothered by corporate competition, fiscal responsibility, or reputation.

Ballistic Background

During the Cold War, the United States and Soviet Russia were locked in a fierce competition for the ultimate high ground of space. The Space Race was thus an international contest that resulted in amazing advances in just a few years. Ballistic missiles allowed this to happen. Let us take a quick look at how these events unfolded, before examining several ballistic spaceplane ideas.

On October 4, 1957, Russia launched *Sputnik*, the world's first artificial satellite. This sent shock waves around the world, and the United States was caught off guard. How could a communist backwater such as Russia have gotten ahead of the West?

Following World War II, German rocket scientists and V-2 rockets were exported to both the United States and the Soviet Union. Neither side had anything nearly as advanced, and so German rocket technology was highly prized. The V-2 rocket became, in effect, the starting line for both sides in the coming Space Race. Benefited by communist secrecy, Russia concentrated on rocket designs able to lift its heavier, less-developed nuclear bombs. The United States, by contrast, conducted its V-2 research program in a relatively open manner. Its own nuclear bombs weighed less than the Russian versions, and as a result there was less of an impetus to develop more powerful rockets. The curious result was that Russia got a step ahead of America in rocketry, because it was a step behind in bomb-making. A more powerful space launcher was the result. With V-2 as the starting line, the Soviet "space shot" of *Sputnik* became the starting gun, and the Space Race was off and running. Within 4 months, the United States put together a tiny scientific satellite,

M.A. Bentley, *Spaceplanes: From Airport to Spaceport*,
doi:10.1007/978-0-387-76510-5_6, © Springer Science+Business Media, LLC 2009

Fig. 6.1 *Explorer 1*, America's first artificial satellite, being held up by (left to right) Dr. William H. Pickering, Dr. James A. van Allen, and Dr. Wernher von Braun (courtesy NASA)

Explorer 1, placed it atop a Jupiter-C ballistic missile, and launched it on January 31, 1958. The American satellite (Fig. 6.1) immediately proved its worth, discovering the now-famous Van Allen radiation belts girdling the Earth.[1]

The next step was putting humans in Space. The X-15 research aircraft began flying in 1959, but it was never intended to reach orbital velocity. Developing an airplane such as the X-15 into a full-fledged spaceplane would take years, possibly decades. Competition to be the first with a person in space was intense. Again, the quickest and easiest way of achieving this was to put someone on a missile and launch that person ballistically. The projectiles would now be people.

Events were happening quickly now, and the only way for either side to remain in the competition was to rely on ballistic missiles. In 1961, two Russians and two Americans were launched into space. The Soviet Union still had the advantage in rockets, and so it was able to launch its cosmonauts directly into Earth orbit. On the

American side, the new Atlas missile was not quite ready; so the less powerful Redstone was used for the first two manned flights of Project Mercury.

Then President John F. Kennedy upped the ante. Just weeks after Shepard's suborbital flight, he made a speech committing America to the goal of landing humans on the Moon by decade's end. The United States would race Russia all the way to the Moon, and do it within 9 years. What happened next, in rapid succession, was a series of space "firsts."

After the first satellite (1957, USSR) and first human in space (1961, USSR), there followed the first double spaceflight (1962, USSR), the first woman in space (1963, USSR), the first three-man crew in space (1964, USSR), the first computer in space (1965, USA), the first spacewalk (1965, USSR), the first change-of-orbit (1965, USA), the first space rendezvous (1965, USA), the first space docking (1966, USA), the first manned Moon shot (1968, USA), and the first manned Lunar landing (1969, USA). This was followed by the first space station (1971, USSR), the first Moon car (1971, USA), and finally, the first Russian–American joint mission (1975).[2] Twenty years to the day after Russia's Yuri Gagarin became the first spaceman, on April 12, 1981, the United States launched the Space Shuttle (Fig. 6.2), the world's first reusable orbital spacecraft. This rapid progress was made possible by Cold War ballistic missile technology, which grew directly out of the German V-2 program of World War II.

Fig. 6.2 Aerial view of second Space Shuttle launch, clearly showing the ballistic nature of the ascent. This photo was taken by John Young, commander of the first Space Shuttle mission (courtesy NASA)

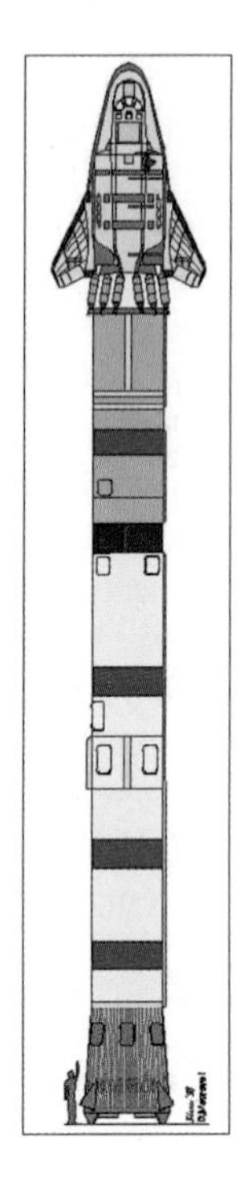

Fig. 6.3 Technical drawing of the Soviet BOR-4 atop its booster rocket, showing the basic launch architecture for Dyna-Soar and other small ballistic spaceplanes (courtesy http://www.buran-energia.com)

The history of manned, as well as unmanned, spaceflight owes nearly everything to ballistic missile technology. And yet, humans were entering suborbital space as early as 1962 in the X-15 spaceplane. And there were other spaceplanes under development as well. The continued and sustained development of these early spaceplanes might well have resulted in a completely different history of manned spaceflight.

In this chapter, we will look at several ballistic spaceplane designs, in which vertical rockets or rocket engines would lob winged vehicles or lifting bodies into space (Fig. 6.3). We will look at the inspirations, the objectives, the designs, and ultimately, the reasons that none of them ever flew. One notable exception, of course, is the most successful spaceplane ever designed, the Space Shuttle. We will look at its own successes and failures in the next chapter.

Dyna-Soar (1957–1963)

The story begins at around the same time as the beginnings of the joint NASA/ Department of Defense X-15 research program. The time was the late 1950s. The project was called Dyna-Soar, short for Dynamic Soarer, and designated the X-20 (Fig. 6.4). Inspired by the original plans of Eugen Sänger and his antipodal bomber, Dyna-Soar was to be a one-person orbital spaceplane capable

Fig. 6.4 Dyna-Soar was planned as a one-man reusable spaceplane, but cost concerns prevented it from realizing its goals (courtesy National Air and Space Museum)

of dropping a bomb on, or conducting reconnaissance over, almost any spot on Earth. It was to be launched by a ballistic missile, returning to Earth as a hypersonic glider, much as the Space Shuttle does today. Unlike the Space Shuttle, Dyna-Soar was to have a relatively compact design, and would balance atop its missile as an eagle on its perch. The launch vehicle was to have been a Titan III missile, then under development for both Dyna-Soar and the US Air Force's Manned Orbiting Laboratory. The idea was to launch it vertically – a method known to work very well – and fly the single-person reconnaissance vehicle back like an airplane, to be launched again another day. There was disagreement about whether to use Dyna-Soar as a spyplane or as an orbital bomber, or both. At various stages of its development, it was touted as the ideal vehicle for each concept. But it never flew. Some preliminary hardware was produced, but stacked up against the two-man Gemini program, its capabilities seemed inadequate, and so the plug was pulled, and Dyna-Soar was canceled on December 10, 1963.[3]

Hermes (1987–1993)

The idea behind a winged, reusable, manned European spaceplane began around 1975 with a French proposal for a smaller version of the US Space Shuttle. This idea eventually became known as *Hermes*, named for the Greek god of boundaries, travelers, and invention. The original plan called for a capability of transporting six astronauts and 4,500 kg of cargo into low Earth orbit.

In the wake of the *Challenger* explosion, it was deemed necessary to add ejection seats in order to bolster crew safety. This, in turn, caused a reduction in capabilities, so that *Hermes* would now carry a crew of three and 3,000 kg into orbit. The loaded launch mass of *Hermes* was 21,000 kg, right at the limit of the Ariane 5 launch vehicle. The spaceplane had a Resource Module for auxiliary systems that would be jettisoned on each flight, just before reentry. A new Resource Module would then be fitted to the reusable spacecraft for the next mission.

But by the early 1990s, Europe and Russia had both signed on to the International Space Station (ISS) project, effectively erasing the need for a manned European spaceplane. International passengers would be able to ride to ISS on either the Russian Soyuz or the American Shuttle. Accordingly, the project essentially fizzled out by the end of 1992 and was officially canceled in 1993. And yet, in the East, there was still hope.[4]

HOPE (1992–2003)

During the early 1990s, Japan's National Space Development Agency studied a small unmanned spaceplane planned for launch in 1999 using its H-II rocket. This H-II Orbiting Plane, or HOPE spaceplane, was to be an 11-m-long, 10-metric-ton unmanned ballistically launched vehicle with a wingspan of 6 m. Its mission was to lift 1 metric ton of cargo to an American space station (which later became the ISS). Later, these specifications were upgraded in a HOPE-X unmanned experimental 13-m-long vehicle with a 9-m span, able to deliver 3 metric tons to ISS. The eventual goal was a manned HOPE spaceplane with a crew of four and a liftoff weight of 22 metric tons.

The original unmanned HOPE and its H-II booster rocket would lift off from the Tanegashima launch center, landing on a conventional runway at another location. As with other spaceplane concepts, HOPE would be an operational, reusable vehicle. It would carry no crew, relying instead on the US Space Shuttle to transport astronauts to and from the space station. The mission of the 10-ton vehicle would be to deliver a relatively small cargo of 1 ton to the Japanese Experimental Module, using the new H-II booster then under development.

From the beginning, it was recognized that this was not the most efficient way to launch cargo into space, since the H-II booster could just as easily deliver a 10-ton payload to low Earth orbit as it could a 10-ton spaceplane with 1 ton of cargo. And yet, this was not the sole motivation behind the concept. HOPE was to use a phased approach in eventually developing an air-breathing spaceplane able to lift itself all the way from a horizontal take-off to orbital space. The real value of HOPE was in its potential as a first-generation reusable vehicle leading eventually to an advanced spaceplane. With the dissolution of NASDA and the origin of JAXA, the HOPE spaceplane project ended by 2004 in favor of other priorities.[5,6]

Kliper (2004–present)

As part of a plan to upgrade and eventually replace the 40-year-old Soyuz design, the Russian firm RKK Energiya has been developing a 6-man winged spacecraft called Kliper. Its spaceplane shape will allow a gliding entry and lower *G* forces than those experienced by Soyuz occupants. The initial proposal came in 2004 and pictured a lifting body with small winglets. Originally, Kliper was to have a doughnut-shaped Service Module with a docking port in the middle, through which crews could gain access to the ISS. Only the forward part, with its lifting-body or winged design, would return to Earth. More recent ideas have the Habitation and Service Module acting as a space-tug named Parom, launched on a separate booster rocket. Kliper would enter space on its own launch vehicle, link up with Parom in orbit, and together they would rendezvous with the ISS. After a mission, the Parom space-tug would remain in space to be used again, while Kliper returned to Earth for reuse. Though highly modularized, this concept retains a fair degree of reusability, which is a definite step in the right direction. But by 2006, with the Russian government rejecting funding for Kliper, the project looked to be in doubt.

In July 2006, Russia and Europe decided to jointly develop a Crew Space Transportation System based on the Soyuz spacecraft. This seemed to doom Kliper, but RKK Energiya still hopes to see Kliper fly by 2012. Will Kliper ever fly?[7,8]

Why No Go?

So there were four small spaceplanes, conceived by four separate nations, each to be launched by a ballistic missile, and not one of them ever flew. In every case, funding was either cut off or never appeared at all. Every one of these spaceplanes was to have perched atop a rocket, ridden into space, and glided back for reuse. Reusability is supposed to be a good thing. Since these were reusable spacecraft, what went wrong?

Dyna-Soar was canceled, in part because NASA's rapidly developing Gemini program was planning to put a two-man crew into space atop a Titan II rocket. This missile, borrowed from the US Air Force, did not require strap-on solid rocket boosters. The Titan III rocket did. And this was the vehicle intended for both Dyna-Soar and the Air Force's Manned Orbiting Laboratory. Dyna-Soar was canceled in favor of Manned Orbiting Laboratory, but it, too, eventually met the dreaded budget axe.

European engineers were concerned about weight as well. The 21-metric-ton *Hermes* was all the Ariane 5 launch vehicle could handle, and much of that weight was in the form of wings, wheels, ejection seats, and the throwaway Resource Module. Downgrading *Hermes* from six to three crewmembers at the same time as its payload shrank from 10,000 lb to around 6,500 lb made the money handlers think twice.

In Japan, the original HOPE concept weighed 10 tons, yet was able to deliver only 1 ton to a space station. And 10 tons was right at the limit of the H-II booster

Fig. 6.5 Spaceflight has always relied on ballistic missiles. Hermann Oberth (foreground), Wernher von Braun (sitting on table), and other officials of America's Army Ballistic Missile Agency (courtesy NASA)

then being developed. The much heavier 22-ton manned spaceplane would have needed a much more powerful booster, such as the proposed H-IID.

Finally, the cash-strapped Russian government was not about to sink money into a tiny spaceplane when it was very familiar with the greater payload capacity of Soyuz and the unmanned Progress supply vessels.

There is a recognizable pattern here, and that pattern has to do with two things: payload capacity and partial reusability. By perching a winged spacecraft on top of a ballistic missile (Fig. 6.5), the wings, the landing gear, the tail, and all the other quintessentially airplane-like characteristics of the spaceplane become dead weight during launch. This dead weight directly subtracts from the payload that the launch vehicle otherwise would be able to place into orbit. As a result, Gemini-Titan outperformed Dyna-Soar, just as Soyuz outdoes Kliper. In addition, the sizes and masses of the reusable components are very small when compared with the sizes and masses of the throwaway components – the huge booster rockets. Therefore, reusability becomes a nonissue. Funding concerns, and other factors such as payload capacity, completely override it.

Winged spaceplanes can be launched ballistically, as we know (Space Shuttle and Buran), but only by using draconian measures. A huge, external supply of propellants, in either liquid or solid form, has to be attached to the reusable spaceplane. Furthermore, these large spaceplanes have relatively large payload capacities, on the order of 30 tons, but again only by using drastically large boosters. For all of these reasons, it seems, the small ballistic spaceplane has never managed to get much past the planning stage. But as we shall see, there are other ways of getting into space, including riding piggyback.

References

1. http://www.nasm.si.edu/exhibitions/gal100/exp1.html
2. Reginald Turnill, *The Observer's Spaceflight Directory*. Frederick Warne, London, 1978.
3. Robert Godwin, ed., *Dyna-Soar: Hypersonic Strategic Weapons System*. Apogee Books, 2003.
4. http://en.wikipedia.org/wiki/Hermes_%28shuttle%29
5. http://www.globalsecurity.org/space/library/report/gao/nsiad92005/part03.htm
6. http://en.wikipedia.org/wiki/HOPE-X
7. http://www.russianspaceweb.com/kliper.html
8. http://en.wikipedia.org/wiki/Kliper

Chapter 7
Piggyback to Orbit

The juvenile stage of spaceplane development involves the cooperation of bigger, more mature rockets. Just as the development of the airplane involved many biwing (and triwing) designs to increase the total lift of the airframe (Fig. 7.1), so the development of the spaceplane has inevitably led to the piggyback concept, and for related reasons. In the case of early airplanes, one wing simply could not generate the necessary lift. Similarly, for current spaceplanes to have any hope of reaching orbital speed, they must be given a lift by a first stage of some kind.

One way of accomplishing this is by perching a small spaceplane atop a ballistic booster, as we have already seen. But another way is to use a booster stage with a winged orbiter mounted to its side or astride its back. This concept takes several forms, including the familiar Space Shuttle configuration. Other ideas usually involve fully reusable systems with manned fly-back boosters. These may take off either horizontally or vertically.

This is where we are today. The most advanced flying vehicle in the first decade of the twenty-first century, in terms of performance, is a piggyback spaceplane. It is a necessary, and painful, step on the road to eventual maturity, as spaceplanes try to grow up.

There are two major modes within the two-stage-to-orbit spaceplane concept, both of which use airplane-like landings. These are the vertical takeoff horizontal landing (VTHL) and horizontal takeoff horizontal landing (HTHL) plans. The Space Shuttle is a partially reusable VTHL space transportation system with four major components, a compromise forced on the space agency by budget constraints. The preferred plan would have had only two components, a manned fly-back booster and winged orbiter. Bristol Spaceplanes have a plan for a TSTO vehicle using an HTHL approach. A huge booster and piggyback orbiter would take off together from a spaceport runway, using the booster's engines to reach as high an altitude and as high a speed as possible. The much smaller orbiter would then separate, ignite its own rocket engines, and continue to accelerate to orbital velocity. This plan is presently only on paper, and it remains to be proven if it can be made practical. As of this writing, the only proven design for a manned piggyback spaceplane is NASA's Space Shuttle.

M.A. Bentley, *Spaceplanes: From Airport to Spaceport*,
doi:10.1007/978-0-387-76510-5_7, © Springer Science+Business Media, LLC 2009

Fig. 7.1 Early airplanes used performance-enhancement devices such as multiple wings to increase lift and aerodynamic cowlings to decrease drag. This is the Curtis Hawk with NACA cowling (courtesy NASA)

The Space Shuttle

The premier example of the piggyback ride to orbit, and one which has been demonstrated over 120 times, is the Space Shuttle. Let us take a look at where the idea came from, how successful it has been, and where it might lead. As noted earlier, the Space Shuttle represents the vertical takeoff horizontal landing concept rather than the relatively unexplored horizontal takeoff and landing mode. If the Shuttle were equipped with jet engines, it could indeed take off from a runway and operate as a jet airplane. This capability was demonstrated many times by the Russian Buran test vehicle, which was fitted with jet engines for this purpose.

The original plan for the Space Shuttle used a huge winged booster that would be flown back to Cape Canaveral after every launch. The manned orbiter would ride piggyback on the booster, and both would lift off from a launch pad much like the Shuttle does today. While the booster's crew was returning to the spaceport, the orbiter's crew would continue into space. There would be no drop-off stages or throwaway tanks. This was to be a fully reusable system, as originally conceived. Owing to cost concerns, the fully reusable concept was vetoed in favor of the present partially reusable design. Ironically, the less costly system actually wastes far more costly hardware on every launch than the more costly system would ever have done. It also has proven far more costly in terms of vehicle attrition. But the real price has been paid in lost lives.

Developed in the 1970s following the Apollo Moon program, the Space Shuttle is the most important spaceplane to date. Despite the loss of two vehicles – each with a crew of seven – the Shuttle has continued to provide heavy-lift launch capability in a reusable winged spacecraft. It is the Shuttle that has delivered, or is scheduled

Fig. 7.2 The first Space Shuttle, *Columbia*, before its maiden flight on April 12, 1981 (courtesy NASA)

to deliver, 27 of the 31 planned major components to the International Space Station. The other four components are being sent, or have been sent already, by Russian rocket. Every Shuttle flight has returned to a precision landing on runways at the Kennedy Space Center in Florida, Edwards Air Force Base, California, or White Sands, New Mexico. Of all the spaceplanes that have been conceived, only the Space Shuttle has proven itself capable of regular orbital spaceflight operations.

The Space Shuttle "stack" is actually a cluster of four components specifically designed to break apart on the way into space (Fig. 7.2). The largest and heaviest of these pieces is the huge bullet-shaped external tank (ET) containing the liquid hydrogen and oxygen propellants for the Space Shuttle main engines. This tank has no engines of its own, but serves as the structural core of the launch configuration. Attached to the tank is the double-delta-winged orbiter and two huge solid rocket boosters, the largest of their kind ever built. The two SRBs burn a solid fuel and oxidizer propellant in a rubber matrix for 2 min, giving a terrific boost to the stack in the initial part of the ride to space. Once they are ignited, they cannot be shut off under any circumstances. When their propellants are expended, the SRBs are jettisoned and recovered in the ocean for refurbishment and reuse. Meanwhile, the orbiter and tank continue to accelerate toward orbital velocity, with liquid propellants feeding into the three Space Shuttle main engines. When its propellants have been expended, the ET is also jettisoned so that it follows a ballistic trajectory and impacts in the Indian or Pacific Ocean. The Shuttle orbiter coasts to apogee and uses its orbital maneuvering system engines to insert the Shuttle into orbit.

By employing powerful solid rocket boosters and carrying the bulk of its liquid propellants in a throwaway gas tank, the Shuttle system can haul large and heavy pieces of hardware into low Earth orbit. The Space Shuttle is perfectly suited to serve as the space truck to deliver space station modules to the International Space Station. The seven crew members provide the human element, ensuring that Space Station assembly is performed professionally and proficiently. Space Shuttle crews have also performed important maintenance on the Hubble Space Telescope, including installing corrective optics. And they have launched probes into deep space, including *Magellan*, *Ulysses*, and *Galileo*.

How reliable is the Shuttle as a launch vehicle? In 120 missions, we have lost two orbiters. This translates to a loss rate of 1 in 60, or 1.7%, which is typical of unmanned space launch vehicles. The remaining Shuttles – *Discovery*, *Atlantis*, and *Endeavour* – are due for retirement in the year 2010 upon completion of the International Space Station. But they still have another 20 or so missions to perform. If the statistical loss rate of 1.7% applies to these 20 missions – and everyone hopes it does not – then there is a 33% probability that we will lose another Shuttle before the program ends. If the unthinkable happens and another Shuttle is lost, then the program will be finished at that point. The remaining two Shuttles will be grounded permanently, and America's manned space program will have to wait until the *Ares* launch vehicle and *Orion* spacecraft are ready to fly, sometime around the years 2012–2014.

To the public, the Space Shuttle is an operational "man-rated" spaceship. Despite the loss of two orbiters, it is deemed safe enough to fly every few months with crews of seven. Three spaceworthy orbiters remain after the losses of *Challenger* in 1986 and *Columbia* in 2003. Before looking at each of these incidents, the alert reader would do well to understand that the Space Shuttle has made far fewer flights in a quarter century than the X-15 made in 9 years. Each Shuttle flies twice a year, on average. The X-15 flew 199 times, but was never considered an operational vehicle. It was a research aircraft only, flown by professional test pilots only. Two of its pilots were Neil Armstrong and Joseph Engle; both later became NASA astronauts. Astronaut Engle later flew the only completely manual reentry in the Space Shuttle. Let us now take a look at what happened to *Challenger* and *Columbia*, and see what lessons we can learn from these events.

Challenger

On January 28, 1986, 1 min and 13 s after liftoff, Space Shuttle *Challenger* exploded. The crew of seven was lost, but their sacrifice was not in vain, because the lessons we can learn from this event will make the spaceplanes of tomorrow

Fig. 7.3 Ice at the pad: cold temperatures on the morning of STS-51-L's launch contributed to the loss of *Challenger* (courtesy NASA)

safe enough to transport millions. This is often the path of progress and is no different than the many sacrifices made by the early pioneers of rocketry and aviation. Only by making mistakes can we learn and improve.

What happened to *Challenger*? Why did it explode? Could it have been prevented? The investigation following the accident revealed that a joint in one of the solid rocket boosters had failed to seal properly. This condition was brought about by a combination of poor design and cold temperatures (Fig. 7.3) preceding launch. A brittle O-ring seal in SRB joint allowed hot combustion gases to escape and act like a blowtorch, aimed directly at the ET. Eventually the tank wall burned through, and the liquid propellants inside were ignited. The result was an immediate fireball, with no chance of escape for the crew.

Solid-propellant rockets work differently than liquid rockets. When a solid rocket is ignited, the propellant grain burns up and down its entire length from the inside toward the outside. The innards of an entire solid rocket case are therefore under continuous high pressure, and the exhaust gases are allowed to escape only through a nozzle at one end, creating the thrust. In the case of *Challenger*, some of these gases escaped sideways through the faulty seal, eventually igniting the liquid propellants in the adjoining tank.

The short reason for the *Challenger* tragedy was improperly designed O-ring seals in the joints of the booster rockets, exacerbated by cold temperatures, poor communications, and inadequate NASA management. It was the usual comedy of errors. But was there another reason for this tragedy? Was there an overarching flaw in the entire system? The answer is yes. Recall that the original design for the Shuttle called for a fully reusable two-stage launcher made up of a fly-back booster and a winged orbiter. Both of these components were powered by liquid propellants. If the original design had been funded by the Congress, there might never have been any solid rocket boosters, there might never have been any leaky joints, and there might never have been a *Challenger*-type tragedy.

What lessons for future spaceplane design can we glean from *Challenger*? The main lesson is that solid rocket boosters should not be used with manned spaceplanes. Clustered launch vehicle "stacks" should be avoided as well. Complete reusability is far preferable to partial reusability, both in terms of safety and operational cost. In the case of *Challenger*, a fully reusable vehicle would have been far safer and less costly in the long run.

Columbia

On February 1, 2003, Space Shuttle *Columbia* disintegrated during reentry, after a successful microgravity and Earth research mission. A gouge in the left wing from a piece of falling foam insulation from the ET allowed hot air to enter the spaceplane, weaken its structure, and lead to its loss. The gouge happened during the launch of *Columbia* nearly 16 days earlier. NASA management was aware of it, but the crew was assured that it would not be a problem.

How could a piece of falling foam cause so much damage? Surprisingly, a heavier piece of debris – such as ice – would have caused less damage than the lightweight foam. How is this possible? When the foam separated from the ET, it entered the slipstream, which quickly *decelerated* it much more so than would be the case with a denser piece of debris. All this happened while the winged orbiter was *accelerating* upward into the foam, which by now had a much larger velocity *relative to the wing* than would be the case if it had been heavier. If the foam had had a greater density and a higher mass, then it would not have slowed down nearly so much in the slipstream, and its speed relative to the vehicle would have been much less, thereby doing less damage. But since the lightweight foam was decelerated so much by the onrushing air after it came loose, both its relative velocity and kinetic energy upon impact with the wing were greatly increased. It was therefore lightweight foam, combined with vertical launch and nonrecognition of the seriousness of the gouge, that led to the loss of *Columbia* and her crew.[1]

Lessons Learned

These are the immediate reasons for the second sacrifice of the Space Shuttle program. But again, was there an overarching reason that caused this disaster? And was there an overall reason for the losses of both spaceplanes? The answer, to both questions, is yes. As with *Challenger*, the clustering of the stack led to the foam impact. Vertical launch of spaceplanes is not a good idea, especially when clustering of the stack is used, because it creates bad interactions between major components. This happened to both *Challenger* and *Columbia*, and it happened during launch in both cases. Paraphrasing Murphy's Law, if anything can fall onto a spaceplane at launch, then it will. This incident is no doubt one of the major reasons NASA is returning to conventional launch architecture in the *Orion* spacecraft, with the crew capsule at the top of the *Ares* stack, topped only by a launch abort system.

The lessons to be learned by the designers of spaceplanes should be these:

1. Avoid solid rocket boosters in or near a spaceplane.
2. Avoid clustering large components around a spaceplane.
3. Avoid vertical launches.
4. Use only fully reusable designs.

We owe it to these 14 brave astronauts, who gave their lives in primitive spaceplanes, to follow these simple rules. Rule number two would seem to rule out piggyback spaceplanes altogether. A winged HTHL booster could be construed as a "large component" clustered around a spaceplane. And this is true. Piggyback spaceplanes are not the best, for more reasons than one, in my humble opinion.

In principle, a Space Shuttle could be launched directly off the back of its 747 carrier, but in that case it would have to carry its propellants internally, or in small side-mounted ETs. The added weight would likely place a terrific burden on the 747, and the Shuttle would probably not make orbit, in any case, before it ran out of fuel. Its payload capability would also be severely degraded, probably relegating it to a passenger vehicle only.

The logistics involved in piggyback schemes are immense. This is illustrated by the fact that whenever the Shuttle lands at an alternate landing site, and requires a piggyback ride back to Cape Canaveral (Fig. 7.4), the cost to the taxpayer is an extra $1 million. The Shuttle first needs to be carefully lifted onto the back of the 747 using the "mate–demate device," securely attached, and then carefully flown cross-country in a series of hops designed to avoid all hazards. Both air traffic congestion and poor weather must be avoided during the transfer. The crash of a 747 while ferrying a billion-dollar Orbiter would be nothing short of scandalous.

These factors will certainly be considered when designing future two-stage-to-orbit HTHL spaceplanes, such as those favored by Bristol Spaceplanes of Bristol, England. The technical, logistical, and financial challenges in this approach are enormous. Right off the bat, in order to develop a working TSTO spaceplane that operates from a runway, two separate vehicles must be designed, engineered, and built. This immediately doubles the complexity, cost, and risk. Then they need to

NASA's 747 Shuttle Carrier Aircraft with the Space Shuttle Atlantis on top lifts off to begin its ferry flight back to the Kennedy Space Center in Florida.

Fig. 7.4 Space Shuttle *Atlantis* leaves Edwards AFB, California, atop its Boeing 747 carrier on July 1, 2007 (courtesy NASA)

be mated in some manner that allows for operational efficiency. The booster stage has to support the weight not only of itself and its load of propellant, but also of its fully loaded and fueled piggyback plane. Without huge resources of capital and manpower, the likelihood of the piggyback concept being developed in the near future looks grim. We will nevertheless take a close look at the Bristol Spaceplanes design concept, because it is a design that could work, in principle. But first, we have a story of a Slavic snowstorm to tell.

Buran

It would certainly be a mental lurch not to mention at this point the Soviet Union's version of the Space Shuttle, which also rode piggyback into orbit. *Buran*, Russian for "snowstorm," flew only once, on November 15, 1988. This was almost exactly 1 year before the Berlin Wall, that concrete symbol of the Iron Curtain, came crumbling down.

At first glance, the *Buran* launch configuration bears an uncanny resemblance to the US Space Shuttle, with its familiar ET and strap-on boosters (Fig. 7.5). But that is where the similarity ends. The launch vehicle for the *Buran* spaceplane was actually the Energiya liquid-fueled rocket. Unlike the Shuttle, *Buran's* main engines were not

Fig. 7.5 *Buran* spaceplane piggybacking its way to the launch pad atop its Energiya booster. Russian rockets are transported horizontally before erection at the pad (courtesy http://www.buran-energia.com)

located in the tail of the orbiter but at the base of Energiya. Also, the strap-on boosters used liquid propellants rather than the solid propellants used by the Shuttle. A third difference is that *Buran* was to have had built-in turbojets. This means that it would be able to make go-arounds on landing, and ferry itself through the atmosphere rather than relying on a piggyback ride on occasion (Fig. 7.6). Finally, *Buran* was able to fly in a completely automated mode, unmanned, which is what it did on its one and only flight. The Russian spaceplane flew successfully and made a perfect automated landing after two orbits of the Earth.

Unfortunately, lack of funding led to its inevitable demise. Not only was the *Buran* vehicle never finished – launching into orbit unmanned because of an incomplete environmental control system – but storage facilities afterward were evidently inadequate, because *Buran's* hangar roof caved in on May 12, 2002, and the spaceplane was destroyed. Like the Berlin Wall and the empire that built it, *Buran* met a somewhat inglorious end.[2]

Ascender

The Bristol Spaceplane has been named *Ascender* by David Ashford, CEO of Bristol Spaceplanes, Ltd. The original idea for the tourist-class spaceplane is to lift off under its own power, rocket up to 100 km, and land on the same runway it took off from. Passengers will enjoy several minutes of weightlessness from the time the rocket engines cut off until the spaceplane reenters the sensible atmosphere (see Fig. 12.1). This is the free-fall portion of the trajectory; it consists of a coast up to the apogee, followed by a plummet back into the atmosphere. Space tourists will gladly pay handsomely for this experience. This is essentially the same launch plan being developed

Fig. 7.6 *Buran*, mounted to its Antonov transport aircraft. Although *Buran* could be fitted with its own jet engines, the piggyback method was also used (courtesy www.buran-energia.com)

by about a dozen launch companies around the world. The initial target altitude is 100 km, because this is considered the international boundary of space. Anyone exceeding this altitude can call himself or herself an astronaut, and rightly so.

Spacecab and Spacebus

Once the suborbital flights of *Ascender* are successful, Bristol Spaceplanes and others have much more ambitious plans to take passengers all the way to orbit. This is where the piggyback ride comes in (Fig. 7.7). In Spacecab, a large winged booster vehicle with the six-passenger Spacecab on its back takes off from Bristol spaceport. The Concorde-like booster is unable to enter space, but does accelerate the Spacecab to Mach 2, using four turbojets, and then to Mach 4, with two rocket engines. Here, the Spacecab orbiter separates from its booster and ignites its own rocket engines while the booster returns to the spaceport. The orbiter carries a crew of two and has a cabin with a capacity for six passengers or space station crew or a payload of up to 750 kg of cargo. Its blunt swept-back shape reflects the fact that streamlining is not required for flight in space but reduces heating during reentry into the atmosphere. Spacecab's job of getting into orbit is now much easier, launching from the upper atmosphere and a speed of about 4,000 ft/s. Spacecab takes its passengers to an orbiting hotel, probably of an inflatable design currently being developed by Bigelow Aerospace.

Fig. 7.7 By taking off in a horizontal attitude, the carrier aircraft becomes the booster, and the piggyback stage is given a free lift part-way to space. The Spacecab concept involves a much more streamlined design (courtesy NASA)

Spacebus is a further enlargement of Spacecab. It will use turbo-ramjets to accelerate the 88-m booster to Mach 4 and two rockets to take the piggyback duo to Mach 6, where separation occurs. The orbiter is a 50-passenger 34-m spaceplane with a 21-m wingspan and a payload capacity of 5.4 tons or 50 passengers.[3]

References

1. David M. Harland and Ralph D. Lorenz, *Space Systems Failures: Disasters and Rescues of Satellites, Rockets and Space Probes*. Praxis, 2005, p. 44.
2. http://www.buran-energia.com/bourane-buran/bourane-desc.php
3. http://www.bristolspaceplanes.com

Chapter 8
Advanced Propulsion

Advances in aviation are nearly always combined with advances in propulsion technology. Better, more efficient engines allow greater speeds or ranges while consuming less fuel. This continual quest for greater efficiencies extends to spaceflight as well. Space launch vehicles have historically carried all their propellants with them, relying on the staging principle to achieve orbit. Berating the atmosphere during launch, regarding it as some sort of fiendish barrier to be crossed as quickly as possible, they nevertheless embrace it on the return journey, thankful for its free drag, which slows the spacecraft down prior to landing. Without this blanket of air around our planet, manned spaceflight would be far more difficult, as all returning vehicles would have to carry as much propellant to slow down as they now carry to accelerate spaceward in the first place.

It is clear that competent spaceflight begins with efficiently traversing the atmospheric curtain that envelops our planet. With this in mind, let us do a survey of some advanced propulsion concepts that may power spaceplanes of the future.

The Aerospike Engine

The aerospike rocket engine represents the epitome of efficiency in the pure rocket. Although it makes no direct use of the atmosphere, it nevertheless operates efficiently and in harmony with local pressure conditions, including the zero pressure of space. Aerospike engines do not look like rockets at all. They are, in essence, inside-out rockets. The familiar bell nozzle configuration is replaced by a curved ramp with either a linear or a circular shape. A series of small thrust cells is arranged along or around the ramp, onto which the hot gases from those cells are expelled. Or a doughnut-shaped combustion chamber may be used, with a circular slit serving as an annular nozzle. The exhaust gases expand in the region adjacent to the ramp and provide an optimum "altitude compensation" from the ground to the vacuum of space. Exhaust products are allowed to expand optimally between the ramp and the ambient atmosphere, instead of being expected to expand properly inside a fixed nozzle. The ramp takes the place of the nozzle walls and provides structure for the rocket exhaust gases to push against, thereby providing thrust for the launch vehicle. While conventional bell nozzles are inherently either overexpanded or underexpanded

M.A. Bentley, *Spaceplanes: From Airport to Spaceport*,
doi:10.1007/978-0-387-76510-5_8,

under virtually all conditions, the unique inside-out ramp of the aerospike ensures optimum expansion of the rocket exhaust products at all times. Hence, ambient pressure conditions allow continuous altitude compensation and peak efficiency, whether the rocket is operating at sea level or in the vacuum of space. Because of this, the specific impulse of an aerospike engine may reach 430 s or more, significantly better than conventional designs, which yield only around 350 s (Fig. 8.1).[1]

Because of the greater efficiencies built into aerospike engines, both the X-33 testbed and its planned successor, VentureStar, were to have used them. VentureStar would have been the world's first single-stage-to-orbit launch vehicle. It was intended to lift off vertically like the Space Shuttle but without an external tank or strap-on solid propellant rockets. Instead, it would have carried all of its liquid propellants internally. Later, it would land on a runway, again like the Shuttle. It would have been a fully reusable single-stage-to-orbit spaceplane. We will revisit the aerospike concept later in this chapter. But first, let us look at some other ideas that make more direct use of the free air around us.

Ramjets and Scramjets

Ramjets have no moving parts. They are essentially hollow tubes containing an inlet-diffuser, combustion chamber with fuel injectors, and exit nozzle. They produce zero static thrust, and so a ramjet-equipped vehicle cannot taxi or take off

Fig. 8.1 SR-71 Blackbird flight-testing linear aerospike rocket engine (courtesy NASA)

Fig. 8.2 Ramjet missile showing simple tubelike construction (courtesy NASA)

under its own power. Therefore some means must be found to accelerate the vehicle to its operating speed before the ramjet can be started, because it is the forward velocity that provides the ram air (Fig. 8.2).

A typical ramjet is designed specifically for subsonic, low supersonic, or high supersonic operation, up to Mach 5. These three designs are exclusive, and do not typically overlap. They differ mainly in the way the inlet-diffuser and nozzle areas are constructed, which depends on vehicle speed. As ram air enters the inlet, it passes through a diffuser section, which slows down and compresses the air, creating a subsonic pressure barrier. The airflow inside the ramjet is always subsonic, even if the vehicle itself is flying at supersonic speeds. High-pressure fuel is injected and ignited in the combustion chamber, and the resulting hot exhaust gases are accelerated through a nozzle. The ram air pressure barrier ensures that the exhaust escapes in the correct direction. A flame holder, situated just forward of the nozzle, stabilizes combustion. The main advantage of ramjets is that they are simple, reliable, and fast. More important, they do not require an onboard supply of oxidizer, typically the heaviest item in any space launch vehicle. The main disadvantage is that the faster the operating speed of the ramjet, the more it must be accelerated by some independent means before it can be started (Fig. 8.3).[2]

Scramjets, despite their name, can never scramble off a runway under their own power, for they suffer from the same basic limitations as ramjets do. They are, in fact, supersonic combustion ramjets, in which the incoming airflow remains supersonic as it passes through the combustion chamber. This allows the vehicle to fly at much greater speeds, up to Mach 15, while using air as an oxidizer and a propellant. As with ramjets, scramjets have to be accelerated to some specific speed – about Mach 5 in this case – before they will begin to operate. Neither ramjets nor scramjets are very good at accelerating themselves, because of a relatively low

Fig. 8.3 Ramjet technology is not new. This Northrup P-61 Black Widow is flight-testing a subsonic ramjet engine slung beneath (courtesy NASA)

thrust-to-weight ratio combined with large drag forces at their high speeds of operation. Pure rockets have thrust-to-weight ratios of around 60, while those of ramjets and scramjets have values of 2 or 3.[3] This is important, because it means rockets are much better at accelerating a vehicle from the ground to spaceflight speeds – from Mach 0 to Mach 25 – than are ramjets or scramjets. Also, ramjets work only in Earth's atmosphere, whereas rockets work anywhere – in atmospheres with no oxygen as well as in regions with no air.

Do subsonic or supersonic combustion ramjets have a future in advanced spaceplanes? As with other concepts, the idea is to get into orbit as efficiently as possible. This means reducing both drag and gravity losses as much as possible. Ballistic rockets minimize drag by launching vertically and punching through the thickest layers of the atmosphere as quickly as possible. They minimize gravity losses by accelerating quickly to orbital velocity before gravitational forces can erode the vehicle's trajectory. The Space Shuttle takes only 8½ min to reach orbit from a standstill on the launchpad. Its trajectory is specifically tailored to minimize drag by launching vertically (Fig. 6.2) and momentarily throttling down the main engines at "max q" – maximum dynamic pressure. At the same time, it minimizes gravity losses by getting into a desired orbit as quickly as practical and before its propellants run out. If the Shuttle were to keep its liquid-propellant engines throttled down for much of the flight, they would burn for a longer period, but gravity would pull the Shuttle into a lower orbit during the burn. The difference between the orbit it was aiming for and the orbit it ended up in would be the gravity loss.

If that orbit were inside the atmosphere it would quickly decay, and the Shuttle would return to Earth sooner than expected.

How would scramjet-powered vehicles minimize drag and gravity losses? The drag on supersonic combustion ramjets is enormous, which is one reason they do not accelerate as quickly as pure rockets. This drawback is offset by the fact that ramjets do not need to carry their own oxidizers. Furthermore, they use the atmosphere as the main propellant. Air is 23.1% oxygen by mass, with the balance being made up of the inert gas nitrogen and trace amounts of carbon dioxide, water vapor, argon, etc. The bulk of the atmosphere is nitrogen, which serves as some 75% of ramjet propellant, even though it does not contribute one bit to the combustion process. This is a trick up the sleeve of the ramjet: the propulsive potential of the nitrogen-rich atmosphere. So this trick somewhat offsets drag losses, since the ramjet vehicle needs to carry fuel only. Gravity losses are much less of a problem, because ramjets can be integrated with lifting bodies, and aerodynamic lift inherently cancels gravity. As long as such a vehicle can generate lift, gravity losses vanish completely. The price for this lift, of course, is drag, as in any aircraft. These two factors are inseparable, and hypersonic drag is a factor that cannot be ignored. Drag goes up exponentially with velocity, although it declines somewhat with air density. But velocity trumps air density in this case, and even at high altitudes, drag causes reduced performance and severe heating.

Scramjets are specifically designed to operate in the upper atmosphere, at an altitude of 100,000 ft or above. If they run out of air, they cease to operate. And if they have not reached a certain speed, they would not operate at all. These are the greatest drawbacks. Their greatest advantage is that they can use the atmosphere as an oxygen source and as a propellant at hypersonic speeds, and provide lift for the vehicle while doing so. But this opens up a whole new range of problems associated with drag, excessive heat loads, and the inevitable search for new materials and designs that can alleviate those concerns. At this point, scramjets lose their simplicity, because complex methods and materials must be used to keep the vehicle from melting. These measures include using advanced materials and often use active cooling methods, such as circulating cryogenic propellants through the vehicle's skin.

For all its merits, the air-breathing scramjet engine operates best at a constant high speed and high altitude inside the atmosphere. It may well have a future in advanced hypersonic airliners, but its future in spaceplanes is still open to debate. If ramjets are used in future space vehicle designs, the flight into space would consist of four phases: (1) an initial acceleration by nonramjet booster to ramjet speed, (2) ramjet acceleration to scramjet speed, (3) scramjet acceleration to Mach 15 at the top of the atmosphere, and (4) rocket acceleration to orbital velocity. This degree of operational complexity, combined with the design challenges of high drag and thermodynamic loads, makes the development of scramjet-powered spaceplanes extremely ambitious, and certainly costly. As in most aspects of advanced spaceplane design, it is fruitful to compare it to other methods and techniques. First, though, let us look at two examples of advanced scramjet technology, and see what lessons they can teach us.

Orient Express: The X-30

During the late 1980s and early 1990s, the United States was actively searching for a way to reach orbit with an integrated scramjet powered vehicle. This eventually took the form of the X-30, commonly known as the National Aerospace Plane. It used a highly integrated propulsion system and structure to allow hypersonic speeds just inside the atmosphere.

The motivations behind the X-30 were two-pronged. If the required technologies could be matured, then the X-30 would serve as the prototype for not only a hypersonic airliner, but also a single-stage-to-orbit spaceplane. As an airliner, it would amount to the modern day version of the Orient Express, greatly reducing flight times between the world's largest cities. The Orient Express would not enter space per se but would hurtle hypersonically through the upper atmosphere using its integral scramjet engines and hydrogen fuel supply, at a speed between Mach 10 and Mach 15. The shovel-shaped air intake would create a large shock wave to compress the airflow prior to entering the combustion chamber. Likewise, the integral aft-body and nozzle would efficiently expand the exhaust products to the rear. The X-30 was to have been a waverider, effectively riding on its own compression shock wave, like a surfer at the beach. The phenomenon of compression lift, which promised greater lift and lower drag, would allow this. The X-30 had a lifting body design and was expected to endure temperatures of 1,800–3,000°F. The project was eventually abandoned in 1993 because of the huge costs in developing the structural materials to withstand the excessive thermodynamic loads at scramjet operation speeds.[4]

The X-43 Hyper-X

When the National Aerospace Plane program was canceled in 1993, it was realized that a greatly shrunken test vehicle could contain costs. With no pilot, the costs of man-rating such a vehicle could be avoided, and test data would still be highly valuable for the eventual development of a manned vehicle. Thus was born the Hyper-X program and the tiny X-43 hypersonic lifting body.

NASA's unmanned X-43 supersonic combustion ramjet research vehicles were 12 ft long, weighed 3,000 lb, and carried just 2 lb of hydrogen fuel. Because they breathed air, they did not require any onboard oxidizer. They were launched with the help of two prestages, a B-52 mothership and an air-launched rocket to bring the test vehicle up to operating altitude and speed. Three were built, tested, and flown at a cost of $230 million. The first flight, in June 2001, failed because the scramjet's booster rocket, a modified lower stage of a Pegasus rocket, lost control in the transonic region and had to be destroyed, obliterating the research vehicle as well. The second vehicle was flown in March 2004 and was accelerated to operating speed by Pegasus, whereupon it demonstrated scramjet flight at Mach 6.83, or about 5,000 mph for some 10 s. The vehicle then became a hypersonic glider, continuing to collect aerodynamic data as it gradually decelerated and descended.

Its flight path led it to an oceanic ending in the Pacific, where it sank. The third flight took place in November 2004, using a modified X-43 with a goal of flying at about Mach 10. This flight also succeeded, with the vehicle reaching Mach 9.68 or nearly 7,000 mph at an altitude of 109,000 ft. This vehicle also ditched in the Pacific. In both flights, it was the Pegasus rocket that boosted the Hyper-X vehicle to its operational speed and altitude. The tests merely validated the operation of scramjet engines at those speeds, without any appreciable self-induced acceleration. As for the Pacific ditchings, this was part of the planned program. None of the X-43 craft were equipped with landing gear or flotation devices, and so they could not be recovered. The tests were performed over the open ocean in order to ensure the safety of the dry-land population. And of course, the vehicle was unpiloted because of its small size; so there was no one aboard to land it.

The Mach 6.83 flight heated the X-43 to 2,600°F, so carbon–carbon was used on the twin vertical tailplanes. The second flight, at nearly Mach 10, was expected to heat the nose to 3,600°F. Because both the successful test flight vehicles ditched in the ocean, they could not be recovered. Therefore, they could not be examined for the thermodynamic effects at the high speeds they endured. In other words, we do not know how much they melted. Nevertheless, the X-43 test program was considered a success, proving that scramjets work at hypersonic speeds in the upper atmosphere.[5,6]

Air-Augmented Rockets

A relatively simple and effective design is to mount a small rocket engine inside a ramjet-like tube so that ram air is further compressed and then accelerated by the rocket. This method is more efficient than a ramjet alone, can double the specific impulse of a conventional rocket, and reduce the weight of the launch vehicle by half. The main working mass becomes the ram air, rather than the onboard propellants. Sometimes called a ramrocket, the hybrid device is more fuel efficient than either a ramjet or a rocket alone. As early as 1965, the Soviet Union was working on such a device, called *Gnom*, which resulted in a multistage ICBM that weighed only 30 metric tons. This allowed the vehicle to be launched from a mobile platform, with resulting strategic advantage. The air-augmented second stage used a solid-propellant rocket and exhibited a specific impulse of 550 s, double what could be achieved with rocket alone. It boosted *Gnom* from its ram air ignition speed to 1 km/s using 300 kg of solid propellant. The final design involved a 29-metric-ton launch vehicle that would lob a 535-kg nuclear warhead 11,000 km. The air-augmented stage, with a ram airflow of 1200 kg/s, would accelerate the missile from Mach 1.75 to Mach 5.5 in just over a minute on an optimum aerodynamic trajectory, after which two more solid rocket stages would take over.[7] The great advantage in air-augmented rockets is that they can reduce total lift-off mass by over half. The Martin company, as early as 1962, calculated that air-augmented rockets could reduce lift-off weight by 56% with a given payload.[8]

Aerial Propellant Tanking

If an oxidizer-propellant cannot be gotten from the atmosphere, are there other methods of getting off the ground without having to lift the heavy oxidizer? One answer may be as simple as refueling in midair (Fig. 8.4). Military aircraft routinely refuel from aerial tankers, greatly extending the range and capability of their forces. The routine and well-practiced nature of these techniques makes it well-suited to serve the spaceplanes of the future. In this scheme, the refueling actually involves a transfer of liquid oxidizer propellant from an aerial tanker to a spaceplane. The technique is called aerial propellant transfer, or APT.

One design that used APT was *Blackhorse*, a one-man spaceplane studied by the US Air Force in the mid-1990s. It was about the size of an F-16 fighter aircraft, took off from a conventional runway, filled its oxidizer tanks from a KC-135 or KC-17 tanker, and zoomed into space from about 43,000 ft. *Blackhorse* would burn rocket-grade kerosene and hydrogen peroxide (H_2O_2), which is a dense oxidizer similar in many respects to water. It was considered to be a "stage and a half" to orbit vehicle, since it enlisted the help of an aerial tanker. The studies showed that such a vehicle could work feasibly.[9]

Another technique is to liquefy atmospheric air and store it in a propellant tank just before "lighting off" for orbit. The Gryphon spaceplane uses such a concept in its Air Collection and Enrichment System. This allows a spaceplane to take off from the ground without having to lift any oxidizer at all, relieving the loads on the landing gear

Fig. 8.4 KC-135 refueling an F-22 Raptor. Aerial fuel transfer is a perfected and routine technique (courtesy USAF)

and associated structure during takeoff. The craft then flies subsonically to as high an altitude as practical, and gradually fills its oxidizer tank with liquefied air. During this phase of the flight, the spaceplane is flying as a normal aircraft, generating only enough thrust to overcome drag. A high subsonic lift-to-drag ratio is required, because as the oxidizer tanks are filled, the vehicle becomes heavier and heavier. A good L/D helps to maintain good efficiency. Only when the oxidizer tank is full would acceleration to orbital velocity take place, from an altitude of around 50,000 ft.[10]

Liquid Air Cycle Engines

One method of utilizing the atmosphere for propulsion has to do with liquefying the air and using it as an oxidizer-propellant in the engine. By using a cryogenic fuel such as liquid hydrogen, air brought into the vehicle can be rapidly cooled until it condenses as a liquid. This liquefied air would then be massively injected into the combustion chamber, along with the onboard hydrogen, and used to propel the vehicle up to Mach 6 or 7.

The disadvantage of these LACE (liquid air cycle engine) techniques is that it not only requires heavy condenser units, but also works only inside the atmosphere. It also requires a much larger fuel flow than do other air-breathing engines, for complicated reasons having to do with temperature pinch points, latent heats of vaporization, and hydrogen embrittlement, all of which are beyond the scope of this book. LACE engines promise specific impulses of some 800 s, compared to a I_{sp} of 400 s for a pure rocket, or 10,000 s for a conventional subsonic air-breathing turbojet.

Turborockets

Future rocket engines will be combined in some manner with turbomachinery so that they can utilize the atmosphere during ascent to orbit. The great strength of the turbojet is the large static thrust that it produces. Therefore spaceplanes so-equipped would easily be able to taxi around a spaceport and take off under their own power. Turbojets are giant vacuum cleaners, sucking in enormous quantities of ambient air, compressing it and combining it with a relatively small amount of fuel. The exhaust is then jetted out the rear of the engine, usually through a subsonic nozzle, to provide the thrust. Turbojets can achieve a specific impulse of 10,000 s, a value which gradually declines with speed.

Turbofans have a larger frontal area, and use this to transfer bypass air around the compressor and mix it with the exhaust to give added thrust.

In a turbo-ramjet, a turbojet is mounted inside the structure of a ramjet, so that the integrated engine can operate from zero forward speed to ramjet speeds. This tends to reduce the efficiency because of the added weight of the turbojet, which ceases to function on its own at about Mach 3.

Turborockets, or air-turborockets, can reach Mach 5 or 6, and are much lighter than turbojets, with a better thrust-to-weight ratio, but have a low specific thrust at low speeds, and a lower specific impulse than the turbojet. By using elements of well-understood turbojet technology, the safety factors of spaceplanes will inevitably rise to the standards common in commercial aviation.

One special type of turborocket is the SABRE precooled hybrid air-breathing rocket engine, shown in Fig. 8.5, and designed by Reaction Engines Limited to power the Skylon SSTO spaceplane. The SABRE engine includes several components, each vital for the efficient operation of the turborocket: an air inlet cone, turbo-compressor, helium loop, small ramjets, and two rocket engines. An outgrowth of the earlier LACE engine designs, which actually liquefied air and separated out the oxygen component, the SABRE engine cools incoming air to just above the point of liquefaction – the vapor boundary. The reason the air is not liquefied is because in LACE engines, too much hydrogen fuel is required to power the system, and the engines become inefficient as a result. To accomplish this, a continuous liquid helium loop is used, which is itself cooled by onboard liquid hydrogen. The helium is an inert element, and therefore ideally suited for double duty as air precooler and energy supply for the onboard turbo-machinery. By cooling down the incoming air, helium is in turn heated, which gives it thermodynamic energy and the potential to do useful work. This heated helium can then turn the compressors, which further squeeze the frigid air prior to being delivered to the combustion chamber. Since the incoming airflow has already been chilled significantly by the closed helium loop, it poses little danger in terms of melting the compressors and other turbo-machinery inside the SABRE. Therefore the entire design can be made much lighter in weight. This is of vital concern in the design of any rocket engine. The two-mode SABRE turborocket can operate

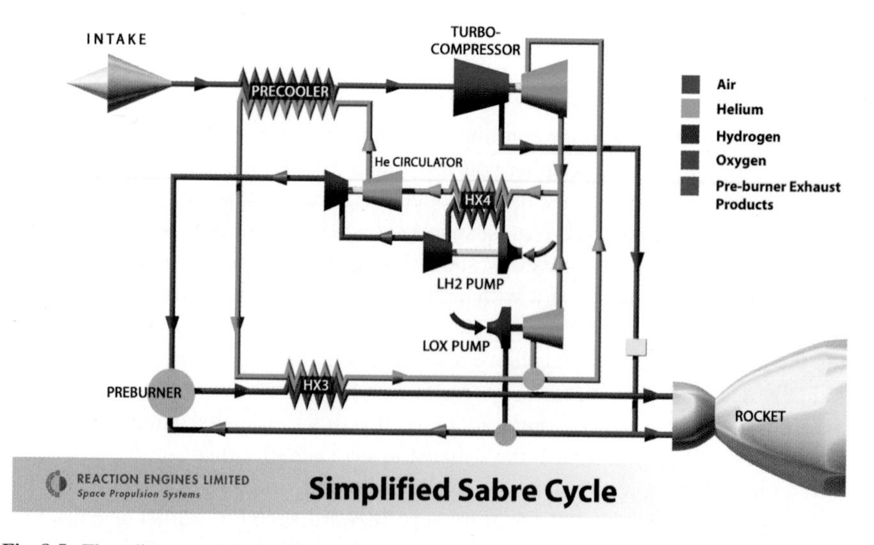

Fig. 8.5 Flow diagram of the SABRE turborocket engine to power the advanced Skylon spaceplane (courtesy Reaction Engines Limited)

as an air-breather from a standstill on the runway up to speeds of Mach 5.5, at which point the rocket engine mode takes over.[11]

Future Design Concepts

Advanced spaceplanes will be powered by air-breathing turborockets of some kind. The specific impulse and delta V (ΔV) advantage of this kind of engine makes it essential. It is useful to keep the rocket equation in mind:

$$\Delta V = I_{sp} g_e \ln R$$

To review, the ΔV, or velocity increment required for spaceflight, depends on rocket exhaust velocity c and mass ratio R. The exhaust velocity is the product of specific impulse I_{sp} and the standard acceleration of gravity g_e, while the mass ratio is the initial fueled mass divided by the final mass of the rocket-powered vehicle after burning all propellants.

An air-breathing engine generates much of its thrust by admitting, compressing, and expelling the available atmosphere. This greatly increases the thrust without increasing the onboard propellant consumption. Specific impulse, you will remember, is the thrust divided by the onboard propellant usage rate. Dimensionally, this is pounds divided by pounds per second, yielding specific impulse in seconds. Because air-breathing engines utilize the atmosphere as propellant, the specific impulse can be as high as 10,000 s, compared to conventional chemical rockets, which have values in the 250–450-s range. The propellant consumption is low in air-breathers, because it involves a fuel only. Most of the working mass used to produce the thrust is the air itself, which also contains the oxidizer.

What features might a future air-breathing turborocket – one capable of powering a spaceplane – include? It will undoubtedly incorporate elements of the aerospike for altitude compensation, some sort of turbine for low-speed operations, and possibly ramjets for high-speed operation. Clever designs and methods will have to be used to keep weights to a minimum.

Imagine several small ramjets mounted tangentially at the perimeter of a turbine, so that when the engine shaft is rotated, the ramjets eventually encounter supersonic conditions, even when the vehicle is parked on the ramp. Spinning up the turbine by pneumatic, electrical, or magnetic means, the ramjets could then be started, to provide continuous rotation of a conventional compressor. In this manner, ramjets could provide the turbine-power for spaceport taxi and low-airspeed operations. While the forward speed of the spaceplane is zero or subsonic, the speed at the turbine tips would be supersonic, ideal for ramjet operation. Since ramjets have no moving parts, this type of engine would be very reliable. Now imagine those same ramjets mounted on pivots that could be rotated forward for supersonic flight. By simultaneously rotating the several ramjets forward, slowing and then halting the shaft, supersonic operations could then be sustained in a nonrotating air-breathing engine with no moving parts. The engine is the same. It has simply changed configuration. The ramjets are now "looking forward." They can now take the spaceplane up to around Mach 5, or about 5,000 ft/s, before the actual rocket engines are powered up.

Spaceplanes by the Numbers

Continuing this line of thought, assume the spaceplane has reached an altitude of nearly 100,000 ft and a speed of well over 3,000 mph. As the air begins to thin, the aerospike rocket is gradually brought online. This uses the same combustion chamber as the ramjet, keeping weights to a minimum. By the time the spaceplane has reached 4,000 mph and 150,000 ft, it is operating on rocket engines alone. It is now up to the rockets to take the ship to 17,500 mph and 528,000 ft. This translates into about a 25,000 ft/s ΔV, including 4,000 ft/s for remaining drag and gravity losses, and 1,000 ft/s as a hedge factor in case earlier assumptions are off. Normal ground-launched rockets require about 30,000 ft/s ΔV to reach orbit; so we have gained a ΔV advantage of 5,000 ft/s, or 3,400 mph with these air-breathing techniques. This might not seem like a lot, but let us look at the numbers, beginning with the exponential form of the rocket equation.

$$R = e^{\Delta V / I_{sp} g_e}$$

With this formula, it is possible to calculate the required mass ratio R for a given ΔV and specific impulse. We will use a I_{sp} of 450 s, yielding an effective exhaust velocity c of 14,480 ft/s. We will also try some other numbers, just for the fun of it. Take a peek at Table 8.1, and then we will discuss what it all really means.

The first three rows show the trend at constant specific impulse as ΔV is reduced from 30,000 to 20,000 ft/s. Exhaust velocity depends directly on specific impulse, and so it remains the same also. Notice the right column. By reducing ΔV by one third, mass ratio is reduced by almost one half. In other words, there is a huge advantage in gaining as much airspeed as possible before igniting the rocket engines. Building a spaceplane with a mass ratio of 4 is a lot easier than building one with a mass ratio of 8. But since we earlier decided on a ΔV requirement of 25,000 ft/s, we will settle on the $R = 5.62$ design. That is still a lot easier than the $R = 7.94$ case.

Table 8.1 Required mass ratios for various ΔV at various I_{sp}

ΔV (ft/s)	I_{sp} (s)	Exhaust velocity	Mass ratio
30,000	450	14,480	7.94
25,000	450	14,480	5.62
20,000	450	14,480	3.98
25,000	400	12,870	6.98
25,000	350	11,260	9.21

The last two rows show the importance of specific impulse and exhaust velocity. As these values are lowered, required mass ratio shoots up. Although it might *just* be possible to build a spaceplane with a mass ratio of 7, building one with $R = 9.21$ would be very difficult, because 89.1% of the light-off weight would have to be propellants. This leaves only 10.9% for ship and payload. Aiming for a mass ratio of 5.62, we find that the propellant mass fraction at light-off has declined to 462/5.62 = 82.2%, leaving 17.8% for the ship and its payload. Let us work the numbers and see if they hang together. Assume a gross light-off weight (GLOW) of 1,000,000 lb – a nice round number. Of this, 822,000 lb are propellants – liquid oxygen and liquid hydrogen. The specific impulse will be an assumed 450 s in efficient aerospike engines. The burn-out weight is 178,000 lb, which includes the ship, crew, and cargo. Now just plug and chug:

$$\Delta V = I_{sp} g_e \log_e R$$

$$\Delta V = (450 \text{ s})\,(32.174 \text{ ft/s}^2)\, \log_e\,(1{,}000{,}000 \text{ lb}/178{,}000 \text{ lb})$$

$$\Delta V = (14{,}478 \text{ ft/s})\, \log_e\,(5.62)$$

$$\Delta V = 25{,}000 \text{ ft/s}$$

The numbers work. By limiting the engines to 2 *G*'s throughout the entire burn to orbit, the spaceplane will accelerate from 5,000 to 25,000 ft/s in 5 min 11 s. The Space Shuttle takes 8 min 30 s to reach orbit, incurring 3 min 19 s more in gravity losses than our spaceplane will. In actual operations, the initial acceleration will start out at less than 2 *G*'s and wind up at more than 2 *G*'s. But this analysis at least provides a basic comparison.

How do all these numbers compare to the Space Shuttle? The Shuttle system has a gross lift-off weight of 4.5 million pounds, including the 58,000-lb ET with its 1.6 million pounds of propellants, the two SRBs weighing 1.3 million pounds each, and the Shuttle orbiter with a lift-off weight of 240,000 lb, including a 55,000-lb payload. Fully 57.5% of the GLOW is in the SRBs. Another 37.2% is in the filled ET. The liquid oxygen *alone* accounts for 30.7% of the launch weight. The remaining 5.3% is in the fully loaded delta-winged orbiter. The 55,000-lb payload makes up only 1.2% of the total (Table 8.2).

The LH_2 tank is 2.706 times the volume of the LOX tank. The LH_2, at a density of 0.5922 lb/gal, makes up 14.45% by weight, and the LOX, at a density of 9.491

Table 8.2 Space Shuttle major components

Component	Lift-off weight (lb)	Empty weight (lb)	Propellants (lb)
SRBs (1,300,000 lb each)	2,600,000	400,000 (both)	2,200,000 (both)
External tank	1,680,222	58,500	1,621,722
Orbiter	240,000[a]	151,205	33,545[b]
Total	*4,520,222*	*609,705*	*3,855,267*
Payload	55,250		

[a]Includes 55,250-lb payload
[b](240,000 – 55,250 – 151,205) lb = 33,545 lb

Table 8.3 Space Shuttle liquid propellants

Propellant	Weight	Volume	Density	Tank
LOX	1,387,457 lb	146,181 gal	9.491 lb/gal	54.6 × 27.6-ft LOX tank
LH_2	234,265 lb	395,582 gal	0.5922 lb/gal	97.0 × 27.6-ft LH_2 tank
°*Total*	*1,621,722 lb*	*541,763 gal*		*153.8 × 27.6-ft External tank*

lb/gal, accounts for 85.55% by weight of the total propellants (Table 8.3). This information will help us define the parameters of our own spaceplane, using the Space Shuttle as a baseline.

Using the Shuttle propellant data above, and beginning with a total propellant weight of 822,000 lb, we find that we will need 200,570 gal of LH_2 weighing 118,779 lb, and 74,093 gal of LOX weighing 703,221 lb. The task is to efficiently package these propellants into a spaceplane weighing no more than 178,000 lb, including engines, passengers, crew, and cargo. These propellants will take up almost exactly half the volume taken up by the Shuttle's ET propellants, but in exactly the same relative proportions. We do not even consider using solid propellants because of their inefficiency, both in terms of specific impulse and operations. In our spaceplane, the liquid oxygen makes up 70.3% of the 1 million pound light-off weight, 85.5% of total propellant weight, but takes up only 27% of the total 274,663-gal propellant volume. This is because it is 16 times denser than the liquid hydrogen. For its part, the hydrogen makes up 11.9% of the GLOW, 14.5% of total propellant weight, but requires 73% of the propellant tankage due to its low density. At light-off of the rocket engines, 17.8% of the GLOW is in the spaceplane, passengers, crew, and cargo, the other 82.2% being in the propellants.

How does the propellant mass fraction of our spaceplane compare to that of the Shuttle? The lower this number, the better. In the following simple exercise, we show how propellant mass fraction f_p is derived. The starting point is mass ratio R, which is the combined spaceship and propellant masses ($m_s + m_p$) divided by the spaceship mass alone (m_s).

$$R = (m_s + m_p)/m_s$$
$$Rm_s = m_s + m_p$$
$$m_p = m_s (R - 1)$$
$$f_p = m_p/(m_s + m_p)$$
$$f_p = m_s (R - 1)/[m_s + m_s (R - 1)]$$
$$f_p = m_s (R - 1)/Rm_s$$
$$f_p = (R - 1)/R$$

There are two different ways to calculate propellant mass fraction: using mass ratio alone, as just derived, or dividing total propellant weight by gross lift-off (or light-off) weight:

$$f_p = w_p / \text{GLOW}$$

Let us see if we get the same number for our spaceplane. If everything is correct, then we should:

$$f_p = (5.62 - 1)/5.62 = 0.822$$
$$f_p = 822{,}000/1{,}000{,}000 = 0.822$$

So far, so good. But how does this compare to the Space Shuttle?

$$f_p = 3{,}855{,}000/4{,}520{,}000 = 0.8529$$

More good news: our spaceplane has a better (lower) propellant fraction than the Shuttle; so it should have a better (higher) payload fraction as well. We can also work the problem backwards, and figure out the effective mass ratio of the Space Shuttle so that we can compare that to our spaceplane.

$$f_p = (R - 1)/R$$
$$0.8529R = R - 1$$
$$0.1471R = 1$$
$$R = 6.797$$

The effective mass ratio of the Shuttle system is 6.797, significantly higher than our own 5.62. The Shuttle achieves this higher mass ratio, of course, by staging. But it also requires this higher mass ratio because it has to lift a much greater weight off the launchpad.

With the help of the numbers we have, a spaceplane can be designed from scratch. The first step is to just guess at the overall dimensions. We know that the total propellant volume is half of what is carried in the Shuttle's ET. And we know that the ET weighs 58,000 lb empty, while the Space Shuttle weighs about three times as much. This compares favorably to our own unfueled spaceplane, which weighs 178,000 lb with crew, cargo, and passengers. To make the ship economical, it should have a good payload capacity of something like 10–20 tons.

That is a starting point. Engineers must figure out how to package all components, integrate the tanks, structure, thermal protection system, and so on, together in a realistic design. It goes without saying that the fuel and oxidizer tanks should follow the same design as those in the Saturn S-II stage, in which they were efficiently sandwiched together with a narrow honeycomb structure in between. Propellant densities will dictate the overall dimensions, and when the first spaceplane begins service as a tanker, its prime cargo should be water. It is safe, has a high density (8.345 lb/gal), and can be turned into LOX (9.491 lb/gal) and LH_2 (0.5922 lb/gal), the best chemical rocket propellants in the universe. Water's high density means that it does not require huge, heavy tanks. The dual use concept for the passenger-carrying spaceplane should be taken very seriously. The gross takeoff weight will be the GLOW plus the weight of fuel required to take the spaceplane from the runway to light-off altitude.

Efficient Nozzles

Whether incorporated into a rotating turborocket or not, future rocket engines will need to maximize their efficiency. Let us now take a closer look at the aerospike rocket engine, and consider whether it can be combined with other concepts as well.

Conventional bell-shaped nozzles are efficient at one altitude only, because the ambient atmospheric pressure affects the expansion of the rocket exhaust. For peak efficiency, the exhaust products must expand within the bell such that they fill the nozzle and emerge "just right." Nozzles are therefore built for certain conditions. Low-altitude nozzles are relatively short, because the rocket exhaust does not need to expand much to match the surrounding air pressure. High-altitude nozzles are much larger, allowing the escaping gases to expand to the much lower ambient pressures at those altitudes. The exhaust gases perform efficient work in pushing the vehicle forward as long as they maintain the proper expansion inside the nozzle. And yet these nozzles are continually afflicted with inefficiencies at every altitude but one, because the ambient conditions are continuously changing during the boost to orbit. They are either "overexpanded" at low altitude or "underexpanded" at high altitude. In overexpansion, the nozzle is too big for the ambient pressure, allowing the exhaust gases to separate from the nozzle walls before they emerge from the nozzle exit plane. In underexpansion, the nozzle is too small at high altitudes or in space, and the exhaust gases therefore balloon out behind the nozzle. Both conditions reduce the efficiency and the specific impulse of a rocket engine.

Aerospike engines solve the problems inherent in conventional convergent–divergent thrust chambers. They do this by turning the nozzle inside-out, so that the expanding gases are directed along a ramp between the atmosphere and the ramp. This results in automatic and continuous altitude compensation, and superior efficiency at all altitudes, including the vacuum of space. Aerospike designs can be either linear or annular. The VentureStar spaceplane was slated to use linear aerospikes to enhance its ability to reach low Earth orbit with a single stage. These engines were considered a vital design feature of the SSTO VentureStar and its unmanned testbed, the X-33.

The annular aerospike engine has "thrust cells" arranged in a circle around a roughly cone-shaped truncated ramp. By angling these thrust cells or using canted vanes in the exhaust similar to those originally used by William Hale over a century ago, the entire engine or parts of it can be rotated. In the Hale rocket, the entire rocket was spin-stabilized using this technique. In the annular aerospike, only certain parts of the engine need to spin, notably the air intake and compressor sections for flight within the atmosphere. The idea is to use a single engine functioning more like a jet in the atmosphere and as a pure rocket in space. This reduces the weight of the spaceplane when compared with designs that use separate jet and rocket engines.

For takeoff and "spike-jet" operations only a portion of the rocket engines need to be used, or alternatively, only a portion of the small thrust cells within the annular aerospikes need to be activated. The number in operation would depend on the total power required to operate essentially as a conventional jet aircraft or to liquefy ambient air for later use during the ascent to orbit. These power requirements are very low as compared with those necessary to achieve spaceflight. As in conventional jet aircraft, the main function of the engines is to overcome drag, once the climb to altitude has been accomplished. The thrust-to-weight ratio during this part of the flight can be less than 1, because the wings are providing the lift aerodynamically, reducing gravity losses to zero.

An Elementary Exercise in Advanced Propulsion

Following is an exposition of how one might go about designing an advanced engine for a single-stage-to-orbit spaceplane. These ideas are admittedly speculative and elementary. Nevertheless, this section may provide food for thought, if not practical plans for future design.

Air-Breathing Centrifugal Rocket Engine

The components of this engine are an air intake, constant speed fan, diffuser section, low pressure seal, centrifugal propellant pump assembly, toroidal combustion chamber (TCC), and truncated aerospike ramp.

The idea is to admit air through the constant-speed fan, compress it to over 300 psi, and inject it into the TCC, where it is burned with LH_2 and ejected onto the aerospike ramp.

The liquid hydrogen and oxygen are pressurized by rotation of the engine. The incoming subsonic air is pressurized by a diffuser section, and supersonic ram air is pressurized by ram air pressure.

The purpose of the diffuser section is to slow down and increase the pressure of ambient air prior to injection in the TCC. This will also raise its temperature, which will increase the combustion temperature and pressure, thus increasing the engine power.

The variable-pitch vanes in the constant-speed fan are necessary to minimize the drag of incoming air and increase the efficiency of the fan. This is the same idea as a constant-speed prop on some airplanes.

LH_2 and LOX are delivered through "spokes" in the centrifugal propellant pump assembly. Air is delivered through an annular section containing a diffuser to slow down and pressurize the air. The air intake admits air at all speeds, from zero (Harrier-type takeoff and landing) to Mach 25.

Impeller-Diverter Rotating Rocket

The components of this engine are an air intake, a hinged air diverter, an annular ram air conduit, a low pressure propellant seal, a centrifugal impeller with integral propellant lines, a toroidal combustion chamber (TCC), and a truncated aerospike ramp. The aerospike ramp provides altitude compensation at all altitudes, and the centrifugal impeller delivers subsonic air, fuel, and oxidizer to the TCC.

During hovering and subsonic flight, the centrifugal impeller sucks air into the engine and flings it outward toward the TCC, compressing it in the process. High-pressure air and centrifugally pressurized liquid hydrogen is then injected into the TCC for combustion and expansion onto the aerospike ramp.

At supersonic speeds, ram air is diverted directly to the TCC, bypassing the impeller completely by means of the air diverter(s). Ram action provides the necessary compression before injection into the chamber. LH_2 is still injected through the impeller assembly by continuous rotation of the engine, which spins in air as well as in space.

Rotation of the engine is accomplished by small thrust cells within the TCC. The proper angular speed of the engine is maintained by controlling the thrust or angle of these cells. This, in turn, controls propellant pressurization and main engine thrust levels.

All propellants except ram air are delivered to the combustion chamber through the centrifugal impeller. These propellants include subsonic air, fuel, and liquid oxidizer, each of which is pressurized by the impeller. Liquid propellants are delivered to the impeller through a low pressure seal located at the axis of the rotating engine, where centrifugal forces cannot disturb the flow of low-pressure propellants. Fuel and (if required) LOX then flow through internal tubes in the impeller, where they are pressurized just before entering the TCC.

One advantage of this design is that the entire engine rotates as a unit. Therefore the only moving part is the engine itself. Rotation accomplishes several things at once: (1) it allows air to be used as a propellant-oxidizer in the atmosphere as it is sucked in and pressurized by the impeller; (2) it acts as a pump for the liquid propellants; (3) it pressurizes the propellants prior to their delivery to the combustion chamber; and (4) it slightly increases the pressure of the ram air because of centrifugal force in the transverse component of the annular ram air conduit. Another advantage of this design is that the impeller is protected from melting at supersonic velocities because ram air is diverted away at those speeds. Yet it still provides pressurized propellants to the combustion chamber by centrifugal operation at all speeds, in the atmosphere and in space. All of these attributes would allow this engine to fulfill its role as a power plant for an advanced spaceplane, incorporating as it does elements of the ramjet, turbojet, and aerospike rocket engine.

References

1. George P. Sutton and Oscar Biblarz, *Rocket Propulsion Elements*, 7th ed., Wiley, New York, 2001.
2. Frederick I. Ordway, James P. Gardner, and Mitchell R. Sharpe, Jr., *Basic Astronautics*. Prentice-Hall, Englewood Cliffs, NJ, 1962.
3. Richard Varvill and Alan Bond, "A comparison of propulsion concepts for SSTO reusable launchers". *JBIS*, 56, pp. 108–117, 2003.

4. http://www.fas.org/irp/mystery/nasp.htm
5. http://www.nasa.gov/home/hqnews/2005/jun/HQ_05_156_X43A_Guinness.html
6. http://www.nasa.gov/missions/research/mach10_meteor.html
7. http://astronautix.com/lvs/gnom.htm
8. http://en.wikipedia.org/wiki/Air-augmented_rocket
9. Captain Mitchell B. Clapp, In-flight propellant transfer spaceplane design and testing considerations. AIAA Paper 95–2955, 1995.
10. http://www.andrews-space.com/content-main.php?subsection=MTA5
11. http://www.reactionengines.co.uk/

Chapter 9
Single Stage to Orbit: The Advanced Spaceplane

Perhaps the reason you are reading this book is because you want to go into space. And you want to enter space in style. You do not want to be lobbed, like a human cannonball. You do not want to ride piggyback on some ballistic behemoth. You do not want to be plastered into your seat like a piece of silly putty. In short, you do not want to be *thrown* into space. You want to be *flown* into space, in first-class comfort. OK, then, read on and find out how your wish may just come true.

Before a child can run, a toddler must walk. Before a toddler can walk, a baby must crawl. And before crawling, the infant must be carried everywhere. During this long period of growth and learning, every child eventually gets to ride piggyback, perching on the shoulders of a parent or clinging to the back of an older sibling. Farm children actually have more fun, riding on real pigs, for example. My brother and I did this. Contrary to popular opinion, pigs are amazingly clean animals when they escape their muddy sties; and they do not mind being ridden by skinny farm kids. They will root up the neighbor's lawn, however, if they should leave the property. We also threw chickens off the barn roof, to test their flight capabilities. These experiments were reasonably successful; the hens made more or less controlled landings despite their less-than-ideal lift-to-weight ratios and their utter lack of prior flight experience. Domestic ducks do better, but you still have to throw them. Ah, but I digress.

The X-15 rocketplane (1959–1968) and the Space Shuttle (1981–2010) are the two most successful spaceplanes to date, with a combined total of more than 300 flights into suborbital space and low Earth orbit. Neither of these vehicles could attain orbit by itself. The X-15 had to be carried aloft (Fig. 9.1) and then crawled into suborbit, while the Space Shuttle has always ridden piggyback to reach orbit.

What are the inherent limitations of these two test programs? In the case of the X-15, the little rocketplane simply did not have enough propellant to reach orbital velocity. Its top speed was 4,520 mph, only a quarter of the required orbital speed of 17,500 mph. Although it exceeded 50 miles altitude on 13 flights and reached 100 km on two of these, the baby spaceplane did so only at the expense of all its forward momentum. The space velocity of the X-15 at apogee was zero every time, having traded speed for altitude. Gravity always brought the little rocketplane back to Earth, converting its potential energy of altitude back into the kinetic energy of high speed. Compared with the mature spaceplanes of the future, the X-15 research aircraft was barely able to crawl into suborbit.

M.A. Bentley, *Spaceplanes: From Airport to Spaceport*,
doi:10.1007/978-0-387-76510-5_9, © Springer Science+Business Media, LLC 2009

Fig. 9.1 The X-15 rocketplane mated to its mothership carrier, the venerable B-52, during a captive flight in 1959. This baby spaceplane never took off under its own power (courtesy NASA)

In the case of the Space Shuttle, it has to cling to the back of another vehicle to have any hope of getting to where it wants to go. Instead of carrying its own rocket propellants, it relies on a huge throwaway external tank and two huge Roman candles – the powerful solid rocket boosters. The X-15 could not carry *enough* fuel, while the Shuttle carries cargo *instead* of fuel, and rides piggyback on its own propellant tanks to get around the problem (Fig. 9.2). The Space Shuttle could be described as a strapping juvenile – too big for its britches – that sometimes gets itself into more trouble than it can handle.

Using the X-15 and the Space Shuttle as benchmarks, is it possible to design a viable single-stage-to-orbit spaceplane that would take off under its own power from a runway and accelerate into space? This problem is not trivial, presenting enormous challenges to the aerospace engineer. In fact, the problem of reaching low Earth orbit from the ground with a fully reusable one-piece spaceship is likely the most difficult problem in astronautics. To fully appreciate the problem, it is useful to look at the history of conventional aviation development.

From DC-3 to DC-X

Aviation, from an operational standpoint, did not take off overnight. Even though the Wright brothers flew their *Flyer I* no less than four times that first day at Kitty Hawk, it was many years before the general public was willing to risk their necks in the new flying machines. Only after Charles Lindbergh, in the *Spirit of St. Louis*, demonstrated that it was possible to fly nonstop across the Atlantic, did general

Fig. 9.2 The Space Shuttle on its mobile launch platform and crawler-transporter, inching its way toward the launchpad. The Shuttle carries virtually all its propellants in an external liquid fuel tank and two solid rocket boosters (courtesy NASA)

aviation take off in earnest. By that time, the days of the biplane (Fig. 9.3) were essentially over. Powerful new engines had been developed, and they were being mounted on the new airplanes in two's and three's (Fig. 9.5). This factor alone assured the public that it was time to climb aboard. One of these early passenger-carrying airplanes was the Ford Tri-Motor, which had three radial engines and carried ten passengers. Several of these lumbering early-birds are still flying, and appear at places such as the yearly EAA fly-in at Oshkosh, Wisconsin. But they were slow, cramped, and inefficient on both the engineer's and the economist's tally sheet. Something bigger, faster, and better was needed (Fig. 9.4).

Fig. 9.3 Still in the biplane era, the Vought VE-7 was a necessary step on the long road toward the spaceplane. NACA test pilot Paul King dons warm clothing and oxygen mask just before high-altitude test flight in October 1925 (courtesy NASA)

The time was right for the first airliner, the twin-engine Douglas DC-3. Though small by today's standards, it was the perfect airplane for its era, and it quickly proved its value in hauling not only passengers but cargo as well. It became, arguably, the first operationally efficient, profit-motive driven airplane in the history of aviation. These aircraft were built by the thousands, with hundreds still flying to this day. They were even used by the Army air forces during World War II, serving with distinction as the celebrated C-47 Gooney Bird. The famous Berlin airlift used many of these useful aircraft. But what does the ancient history of aviation have to do with the advanced spaceplanes of the future?

Fig. 9.4 The Vega Air Express, with NACA-designed cowling, broke the transcontinental speed record in 1929 (courtesy NASA)

Fig. 9.5 Twin-engine aircraft greatly increased the range and reliability of airplanes in the years before and during World War II. Mercury astronaut Donald K. "Deke" Slayton (*right*) and Ed Steinman (both 1st Lieutenants) pose with their Douglas A-26 bomber in 1945. Astronaut Slayton later took part in the Apollo-Soyuz Test Project in July 1975 (courtesy NASA)

It actually has a lot to do with it, because until airplanes became operationally efficient, they never appeared in large numbers. And until they appeared in large numbers, they did not have a chance of becoming economically operational. The key was people – thousands of fare-paying passengers who simply wanted to go. If they had the money, and enough of them obviously did, then they could get to ride in style on one of the sleekest, fastest ships then available – the DC-3. These airplanes – these ships of the sky – even had captains and flight crews, something which has held over to the modern day. Not everyone, of course, could afford to pay the ticket price, and only the well-to-do took to the skies in the early years. And so it will be with the first spaceplanes.

The DC-3's operational success was linked to its versatility, its reliability, its safety, and of course, its reusability. Every successful passenger airplane since then has taken its cue from this venerable old maiden of the skies. This cannot be said of the DC-X Delta Clipper, which was to have taken this legacy beyond the skies.

The DC-X was one of the first single-stage-to-orbit VTVL testbeds ever flown. And fly it did, though it never flew in space. It was basically an Earth version of the Apollo Lunar Module, without a crew. It could take off, hover, fly horizontally some distance, land on its four legs, and do it all using rocket power alone. One of its remote control pilots was Charles "Pete" Conrad, Commander of *Apollo 12* and the third man to walk on the Moon. On the last flight of the much ballyhooed Delta Clipper, it tipped over and blew up. And that was that. There are plenty of reasons why this happened, including lack of funding, an overworked ground crew, and so on. But the fact remains that according to Murphy's Law, if something can tip over, it will tip over. In the case of the DC-X, one of its landing gear failed to extend at touchdown because of an unattached hydraulic line. Although it had flown perfectly, and its other three legs had deployed as they should, this one little glitch was enough to doom the top-heavy craft. Toppling to the ground, the remaining onboard propellants immediately ignited, and the spacecraft was destroyed in a blazing instant.

What lessons are involved in this comparison of DC-3 to DC-X? There are many lessons, some so simple that they would tend to be ignored completely. Two of these cannot be ignored here, however. The first lesson is that piloted machines are inherently safer than unpiloted ones. Every DC-3 had a flight crew. If the DC-X had a pilot, you can be sure he or she would have checked the landing gear before taking off. This is normal operating procedure. Pilots are trained never to fly an unsafe craft. The second lesson is that it is very unwise to fly a machine that may tip over, especially if tipping over means blowing up. Airplanes cannot tip over, because they have wings sticking out on both sides. Even if a landing gear collapses completely, the airplane merely tilts a little and slumps to the tarmac (Fig. 9.7). Simple, but who would think of it? Who would speak of it? It is a silly thing, a simple thing, but it is true. *Airplanes do not tip over!* And neither will spaceplanes.

Before we get into the what, the why, and the how of the advanced spaceplane, it is convenient at this juncture to note that one of the reasons the X-33 program was canceled was because of the concerns of flying unpiloted craft over the continental United States. The X-33 was to have tested the new linear aerospike rocket

Fig. 9.6 Douglas D-558-2 Skyrocket in flight with F-86 chase plane, during the mid-1950s. The post-war years saw a rapid and continual improvement in airplane and rocketplane designs, exemplified here by the recently introduced swept wings of both aircraft (courtesy NASA)

Fig. 9.7 Rocketplanes might trip up, but they do not tip over. Here, the nose gear of the X-1E has collapsed on landing at Rogers Dry Lake in June 1956. The research aircraft was easily repaired to fly again. See also Fig. 2.5 (courtesy NASA)

engines on suborbital trajectories, lifting off from the American southwest and landing on runways in Nevada and Montana. But the specter of one of these unmanned vehicles crashing – and causing loss of life or injury to those on the ground – was too great. This ultimately contributed to the cancellation of X-33 and its follow-on project, VentureStar (see Fig. 9.8).

Fig. 9.8 The X-34 Technology Testbed Demonstrator spaceplane parked on the ramp at Dryden Flight Research Center, California. Like the X-33, this promising program was also unfortunately canceled (courtesy NASA)

The Spacefaring Public

Who will ride on the advanced spaceplanes of the future? Anyone with enough money who wants to go, and anyone with a destination in space. This will include space tourists, astronauts commuting to space stations, and inanimate cargo. The very same factors that funded the phenomenal success of the DC-3 will be responsible for ensuring the success of the advanced spaceplane, as well as the suborbital spaceplanes that precede it. If the public feels assured that the vehicle and its engines are reliable and safe enough, they will line up by the thousands to buy their tickets. This is already happening, albeit on a much smaller scale. With the success of SpaceShipOne in 2004, and the announcement by Virgin Galactic that they would soon begin offering suborbital space rides, lines quickly formed. Several hundred space tourists have already put down payments on future space experience flights. And this is before any real destinations have even been set up yet. This may be just as well, since it will be a while before spaceplanes have matured to the point of reaching orbital speed. Yet the fact that hundreds of future space tourists have already reserved their seats reveals that they will pay for the experience because they want to, not because they have any particular destination in mind. The spaceflight experience is destination enough.

Why would anyone want to ride into space? Because it is there, and because it is different. There is an insatiable urge of the human spirit to experience that which has never been experienced before, to go to places new and exciting. In my own case, I wanted to go to Sweden, a country my forebears had left a hundred years ago. The destination was there; I had a personal connection to it; so I worked for 2 years as a busboy, and finally went, all before reaching the age of 17. I got there by boarding an airliner in Denver, changing airlines in New York (waiting there cheerfully for 11 h), and flying to Europe by way of Iceland. I used the transportation readily at hand, which ultimately included the punctual European railway system. In the case of my ancestors, they boarded a ship somewhere on the coast of Sweden, and spent several weeks making the voyage to Boston. In both cases, we went because we wanted to, no other reason.

What kind of vehicle will space tourists of the future want to ride on? They will have their choice, in the coming years, of ballistic missile or spaceplane. Those with a higher degree of derring-do will doubtless choose the rocket, especially if they are as financially free as they are fearless. Those with less financial fortitude will choose the spaceplane, which will still be in an immature, suborbital stage of its development. These events will have important consequences for the future, because the spaceplane will garner far more customers than will the missiles, even if they do not go as far, as high, or as fast. Most important, however, is the inevitable result that those companies who operate spaceplanes will gain far more experience in spaceflight than those who concentrate on rockets alone. Both missiles and spaceplanes use rocket engines, and rocket engine experience is every bit as important as spaceflight experience. Furthermore, the most valuable flight experience in a spaceplane happens while it is flying through the atmosphere, not while it is floating in space.

As spaceplane companies begin to turn a profit with their suborbital clientele, there will be a continual push for ever faster craft that can reach ever greater altitudes, and offer their passengers better views for slightly longer periods of time in weightless conditions. While missiles are blasting into the heavens perhaps on a monthly basis, spaceplanes will be soaring into suborbit on an almost daily basis. Like the tortoise and the hare, those who follow the logical path of development from the suborbital spaceplane to the fully orbital article will win the race. Those who get side-tracked by ballistic or two-stage-to-orbit schemes will ultimately lose when it comes to space tourism. Given a choice between riding piggyback to orbit, or first class to suborbit, which would you choose, especially if the piggyback ride cost you 100 times as much?

As I write these words, space tourists are routinely expected to pay as much as $30 million to ride Russian rockets to the International Space Station. Virgin Galactic, meanwhile, expects its passengers to pay less than 1% of this figure, something on the order of $200,000. It is obvious who will get the most passengers and who will gain the most experience. The best course of development for the spaceplane is one that begins with suborbital craft similar to the Rocketplane XP, SpaceShipOne, or the XCOR Xerus, and ends up at some point in the future with fully capable single-stage-to-orbit designs that take off from runways under their own power and fly into space without the aid of external tanks, motherships, or ballistic boosters. Advanced engines will inevitably enable this kind of future to

Fig. 9.9 Unmanned X-48B parked in desert. Is this the shape of future spaceplanes? (Courtesy NASA)

materialize before our very eyes. Like the tortoise, or the Little Engine That Could, these companies will someday win out over all.

Anatomy of a Spaceplane

The conventional airplane uses the atmosphere to (1) sustain combustion in the power plant, (2) generate lift against gravity, and (3) provide propellant. The advanced spaceplane in our future will do likewise, and will be similar in many respects to a very large jet airplane. There will be important differences, however. It may have a delta wing, or a double-delta wing similar to the Space Shuttle orbiter. Or it may be designed as a lifting body without wings as such, but able to generate lift with its entire fuselage. A third possibility is that it may be long and cylindrical, with pointed ends and short stubby wings. Whatever shape it takes, it will sport advanced air-breathing engines similar to those described in the last chapter. Unlike the Shuttle, the advanced spaceplane will carry far less cargo but far more propellant. Its main function will be to transport spacefaring tourists into low Earth orbit and to the Moon. As already indicated above, the advanced spaceplane will not rely on booster rockets, external propellant tanks, or carrier aircraft. Earlier models may engage in some sort of aerial refueling, but eventually even this necessity will be eliminated by the proper application of optimal fuel and flight management, lightweight structures, and improvements in engine technology.

The advanced spaceplane will be able to fly into Earth orbit on one tank of "gas," refuel in orbit, and take its passengers to the Moon. In time, it will have the ability to land at a Lunar resort, collect lunar-derived propellants, and return to low Earth orbit with both propellants and passengers. Using its winged configuration, it will aerobrake in Earth's atmosphere before rendezvousing with an Earth orbital propellant depot. Serving double-duty as a spaceliner and space tanker, the advanced spaceliner will fill a vital niche in the space infrastructure of the future. Spacelines will make money hauling not only passengers but propellants as well. The constant flow of propellants from the Moon, and passengers from Earth, will fuel the space economy, dwarfing anything we see today.

The spacelines of the future will operate much as the airlines of today, with the additional responsibility of transporting cargo – mainly propellants – between the Moon and the Earth. Some spacelines will transport dry goods rather than passengers, but all will transport propellants, because without this continual flow of rocket fuel, the space lanes will remain untraveled.

But what unique characteristics will allow the advanced spaceliner to accomplish all of this? The single most important factor will be advanced propulsion technology. As the Wright brothers knew, a good engine is just as important as a good airframe. Just as they had to build their own engines, those who today are building their own engines bear close scrutiny, for they will likely be among the major players in the space economy of tomorrow.

A peek inside the spaceplane reveals that it consists mainly of huge propellant tanks, as shown in Fig. 9.10. These are necessary to propel the vehicle to an orbital velocity of 5 miles/s. This is the biggest difference between the advanced spaceplane and today's airplane. Airplanes carry most of their fuel inside their wings,

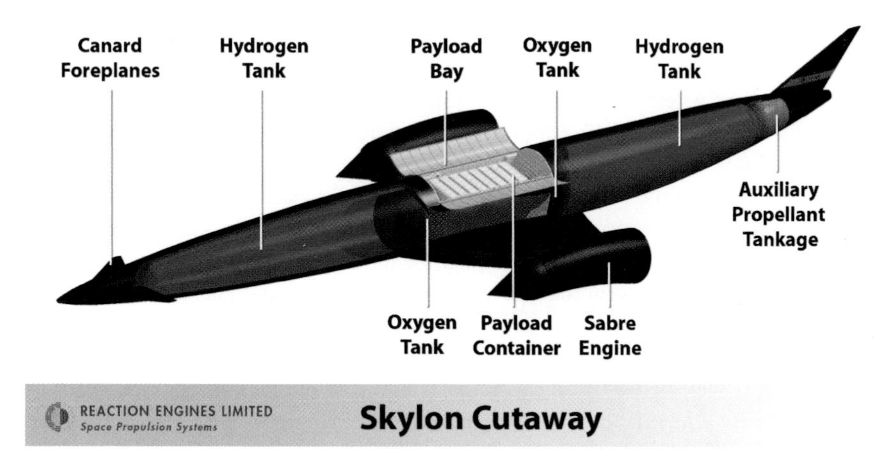

Fig. 9.10 Internal arrangement of the Skylon SSTO spaceplane concept (courtesy Reaction Engines Limited)

Fig. 9.11 There is no real up or down in free-fall, in orbit, or between planets. These astronauts are enjoying a few moments of free-fall inside NASA's converted KC-135 (courtesy NASA)

while the spaceplane must carry a large proportion of it inside the fuselage as well. The body of the spaceplane may have a bulged appearance, so that it can contain enough propellant for it to reach orbital speed. This is a good thing, of course, because the same design attribute which is required to enable SSTO spaceflight also makes the advanced spaceplane the perfect space tanker, which is needed for the Lunar tourist economy.

But what about tourist accommodations? Spaceliners bound for the Moon will be outfitted with a large number of viewports, lounges, sleeping areas with "bunk-bags," and zero-G restrooms. Unlike today's International Space Station, which appears on NASA television to have an untidy, jumbled interior, the spaceliners of the future will arrange their cabins so that there is a "visual up" and a "visual down." There will be decks, bulkheads, and overheads, as in any ship. On the deck will be neatly arranged tables and chairs, on the bulkheads will be port holes and paintings, and in the overhead will be recessed lighting. There may be shelves of books on some walls, with special arrangements to keep them in place. And there will be flat-panel display screens, firmly anchored to the bulkheads, with a definite "up" and "down." If some passengers wish to float upside down during the zero-*G* flight to the Moon or in orbit, they will be free to do so (Fig. 9.11). But for the psychological well-being of all passengers, the accommodations will be made to appear as normal as practical. Above all, the jumbled appearance so characteristic of the interior of government spaceships will be avoided.

What will the cockpit look like? The pilots' stations will look very much like those you would find in any airliner, and many of the controls would be familiar to any pilot. The control yoke will be replaced with a control stick, similar to those found in jet fighters. This will be done for the practical reason of attitude control in space. In flight, aircraft have three mutually perpendicular axes of motion: pitch, yaw, and roll. Pulling back on the stick pitches the nose up, and pushing forward pitches the nose down. Moving the stick to the left rolls the wings counterclockwise, and moving it to the right rolls the wings clockwise, as seen from the pilot's station. Yaw control in the atmosphere is normally accomplished by means of foot pedals, but in space, a twist of the stick will accomplish this function, because the hand is far more dexterous than the foot. Therefore, although the spaceplane will be equipped with foot pedals for yaw control in the atmosphere, this function will be accomplished by a control stick twist while in space. There will be a switch on the instrument panel or pilot's console that will enable or disable attitude control thrusters. At heights above 100,000 ft or 20 miles, this switch will be turned on so that the spaceplane will not lose control due to lack of sufficient aerodynamic forces operating on the conventional control surfaces.

Radar altimeter will be one instrument common to all spaceplanes, especially those outfitted for Lunar flight. Conventional aircraft rely on barometric air pressure to indicate altitude. But because there is no air on the Moon, the only way to accurately determine altitude above the ground is by radar.

The advanced Lunar spaceplane of the future will be equipped with ventral thrusters to enable landing in the manner of a Harrier jump-jet. This maneuver may be termed vertical landing in a horizontal attitude, or VLHA. The same acronym refers to vertical liftoff in a horizontal attitude. Lunar-landing spaceplanes will all be VLHA vehicles. A level landing pad cleared of all debris will be necessary before this is attempted. Unlike the DC-X, there will be no danger of the Lunar spaceplane tipping over on the Moon, for the same reason that airplanes are not plagued by this problem on Earth. Spaceplanes of the future will have VLHA capability on the Moon long before they develop this ability on Earth. The reason, of course, is that the Moon has only one sixth the gravitational acceleration of Earth, and so only one sixth the thrust is required to land or lift off.

Eventually, ultra-advanced spaceplanes will appear, with the ability to perform VLHA on Earth as well as on the Moon. This capability will actually increase the safety of these craft, because they will have a much smaller probability of crashing in bad weather on Earth. They will simply "hover" their way to a precise radar-controlled landing on any planet. These kinds of craft will likely not appear for many years, but they will appear eventually.

Spaceline Timeline

The more advanced of these spaceplanes will take years, and probably decades, to develop. As for the suborbital vehicles, they have already arrived. In fact, they have existed since the X-15 first flew in 1959. The natural maturing process of these suborbital spaceplanes was arrested for almost half a century by an intervening developmental eclipse by ballistic boosters. These "big dumb boosters," for their part, have already reached as mature a stage in their own development as they can, despite the fact that they never have (and probably never will) reached the same standards of safety expected of an airliner or a spaceliner.

When will the advanced spaceplane arrive, then? This is a question no one can answer at present, confounded by the fact that there is no consensus on how to go about developing them. There are many competing visions of the future, each with its own specific vehicle concepts and engine designs. Some envision a future without spaceplanes of any sort, a future of spaceflight that includes DC-X type vehicles only. Others see a future of two-stage-to-orbit spaceplanes as an intermediate step on the road to the mature single-stage-to-orbit spaceplane. Still others see a gradual improvement of suborbital spaceplanes until they achieve orbital spaceflight capability at some point.

Just as the Wright *Flyer* was incapable of transatlantic or supersonic flight, so SpaceShipOne was incapable of orbital or Lunar spaceflight. And yet the supersonic Concorde utilized the very same principles of aeronautics to fly across the Atlantic as the *Flyer* used at Kitty Hawk. By a series of small steps, punctuated by a few huge leaps, aviation progressed from the simple biplanes of yore into the hypersonic research vehicles of today. Similarly, by a continuous and sustained effort, the spaceplanes of today will grow into the advanced spaceliners of tomorrow. This process will accelerate when the first space tourists begin to fly on suborbital spaceplanes, because flight frequencies will increase, and with frequency, an ever-growing experience base.

The first advanced spaceplane, defined as one capable of taking off from a runway under its own power and accelerating into orbit (Fig. 9.12) with or without the

Fig. 9.12 The advanced Skylon spaceplane in a low Earth parking orbit (courtesy Reaction Engines Limited)

help of an aerial tanker, could fly sometime around the year 2030, give or take 10 years. As these words were written in late 2007, it was still very difficult to predict what might happen, given the fact that suborbital space tourism had not yet begun. But when this seminal moment arrives, a chain of events will unfold that will be wholly unlike anything we have seen before.

Orbital space tourism, it should be noted, has already begun. Since 2001, half a dozen space tourists have ridden Russian rockets to the International Space Station. Each of these space tourists has paid $20 million or more for the privilege of a few days in space, rubbing shoulders with professionally trained astronauts and cosmonauts. These flights have been relatively infrequent, and the funds have gone to the Russian government rather than to private spaceline companies. Therefore, they will have virtually no effect on the overall development of space technology, unlike the impact suborbital space tourism is about to have. As noted above, suborbital space tourism will cost the space tourist less than 1% of what the orbital space tourist today pays. This means there will be 100 times as many suborbital tourists, with the potential for much greater flight frequency, spaceflight practice, and spaceplane development. And it is the spaceplanes themselves – including their purchase, operation, and development – that the space tourist will gladly pay for.

Have Spaceplane Will Travel

With the rapid development of the spaceplane – strictly suborbital in its first incarnation – there will be a rush to upgrade ordinary airports to spaceports. This will carry a certain prestige value, which is always good for business. Eventually, spacelines will not only offer space tourist rides, but also offer suborbital hops taking passengers to neighboring states or countries. As with space tourism, the first of these flights will be motivated by the fun factor. They will be joy rides, in essence. Soon, however, people will realize that there is real value in the suborbital flight. Critical materials, such as organs destined for transplant, time-sensitive documents, or diplomatic pouches, will find their way aboard the suborbital spaceplane. The concept of same-day mail could take off. International rescue workers could be quickly spacelifted to the scene of some disaster, in a manner reminiscent of *The Thunderbirds*.

Airports around the globe, at this point, will begin applying for upgrades to the status of spaceport. The first suborbital "range flights" will be relatively short. Spaceplanes will take off from runways in southern California and land in Las Vegas. Others will depart Spaceport America in New Mexico and land within a few hundred miles. Over time, the range of the suborbital spaceplane will increase until it is hopping from continent to continent.

Once these suborbital vehicles have advanced to the status of true SSTO spaceplanes, the first destinations will be "space hotels" in low Earth orbit. These will be inflatable habitats pioneered by companies such as Bigelow Aerospace. The space experience will then be greatly enhanced by the luxurious accommodations awaiting the eager space tourist some 200 miles above Earth. A suite of zero-G activities

Fig. 9.13 The conceptual Russian-designed Spiral spaceplane, shown entering the atmosphere (courtesy http://www.buran-energia.com)

will be available, with possibilities limited only by the imagination. Travelers will have their choice of tour packages, some electing to stay "overnight," which will include a dozen or more sunrises and sunsets, others choosing to stay longer. There will be "quickie cruises" lasting 12 h or eight orbits around Earth, all the way up to weekly or even monthly packages, each priced accordingly. For the return to Earth, travelers will be offered their choice of landing sites as well. In this way, their space vacation may start in America and end in Australia, for example. Spaceplanes, with their large cross-range reentry ground-track capability, will be ideally suited to offer such choices to their customers.

Peering a little further into the future, the Lunar spaceplane will drop off its passengers at an Earth orbital resort while it goes to refuel at a nearby depot. After several hours, the spaceplane will return, pick up its passengers, and "light off" for the Moon. The first Lunar spaceplanes will not land on the Moon, but will instead transfer passengers and cargo to crablike Lunar shuttles (Fig. 9.14) based there permanently. Later, the advanced Lunar spaceplane will demonstrate the ability to land on Luna itself, as we will see in the next chapter. The Lunar tourist will follow closely on the heels of the ordinary orbital tourist, because the Moon offers a whole new set of fun-filled vacation opportunities. When that day arrives, honeymoons will never be the same.

Flank Speed to the Future

How will future spaceplanes utilize their inherent strengths to enter space? Will they assault the problem with brute force, as ballistic rockets do? No! Instead, they will use a flanking maneuver, hitting the problem sideways in a clever, more competent fashion. At the core of every spaceplane, of course, will be its efficient and able engine.

Fig. 9.14 Lunar-based shuttles might resemble this Lunar landing research vehicle, used by Apollo astronauts to practice Moon landings (courtesy NASA)

The key to the advanced spaceplane will be advanced propulsion technology, new, lightweight engines that will propel us into the future. As we have already seen, these engines will be air-breathing hybrids – part turbojet and part rocket – able to function as efficiently in the atmosphere as they do in space. Yet these new engines are only part of the strategy in out-flanking the future. A proper ascent profile through the atmosphere is also required.

There are many ways to enter space, and just as many ways to return. There is the tried and true vertical launch with ballistic landing capsule (Fig. 9.15), demonstrated from 1961 through 1975 in the US manned space program, and soon to be rehabilitated. There is the vertical launch with horizontal landing (Fig. 9.16), as in the Space Shuttle. And there is the horizontal takeoff and horizontal landing, the airplane approach of the advanced spaceplane. In the first two methods, the launch vehicle punches through the atmosphere, then arcs over to build up enough speed to stay in orbit. But what ascent profile might the advanced spaceplane use?

It turns out that the advanced spaceplane's ascent trajectory will be inextricably linked to its onboard power plant – the air-breathing turborocket. Working more

Fig. 9.15 Splashdown of the Apollo 9 Command Module, the original American method of returning spacefarers to Earth (courtesy NASA)

like a turbojet at takeoff and at low altitudes, the advanced spaceplane will depart the runway much like any ordinary airliner. The flight profile in the first part of the ascent to orbit will therefore be very conventional. A small amount of onboard fuel will be burned with a large quantity of air, and this will serve as the working mass during this stage of the flight. The duration of this phase will depend on whether the spaceplane is designed to collect, liquefy, and store air before accelerating into space, whether it "refuels" in mid-air, or whether it has taken off with all the propellants it needs to reach orbit. Either way, at some point the spaceplane will make a relatively rapid acceleration, get above the atmosphere, and enter orbit. It will remain in the atmosphere as long as there is a benefit to doing so, but as soon as the air becomes too much of a "drag," it will be time to get out.

Fig. 9.16 Artist's rendering of the SF-01 lifting off with space tourists aboard (courtesy Spacefleet Ltd.)

The lift generated by the advanced spaceplane will completely cancel gravity losses as long as it operates inside the atmosphere, and drag will be overcome by sleek design and by exiting the atmosphere before the speeds get too high. Thermodynamic – or heat – loads will be kept down the same way as aerodynamic loads, because both arise together. Let us now take an even closer look at the advanced spaceplane's bag of tricks.

Tricks of the Trade

The greatest asset of the advanced spaceplane will be its engines. They will be far more versatile and powerful than anything before, using elements of the turbojet and aerospike rocket in their designs. These attributes will make them, and the spaceplanes they power, very reliable and efficient. As we have seen, engine and vehicle reliability is what is required to make spaceflight routine and safe. You have already read about these engines in the last chapter; so we will not say anything more about them here.

Beyond the value of its engines, the advanced spaceplane will use advanced flight and fuel management (FFM) techniques. Flight and fuel management refer to the proper utilization of the atmosphere for lift, oxygen, and working mass, as well as choosing the optimal path and velocity profile through the atmosphere. By keeping the ascent to orbit as efficient as possible, the least amount of fuel – typically liquid hydrogen – will be needed to reach space velocity. Let us now look at each aspect of FFM to see what information we can glean with current knowledge.

The first and most obvious element of FFM is *aerodynamic lift*. Spaceplanes will use the atmosphere to assist them in ascending to space height. This is fairly straightforward, and the best understood aspect of spaceplane operations. Airplanes use the atmosphere in this way all the time, although every airplane has a ceiling beyond which it cannot climb. Winging its way up to around 50,000 ft, the advanced spaceplane can remain subsonic, flying just like a commercial airliner. An excellent subsonic lift-to-drag ratio is employed to minimize drag and fuel consumption during this stage of the flight. In this way, aerodynamic lift provides much of the altitude that the spaceplane craves, all for the price of a modest amount of fuel.

The next element is *oxygen*, required for the operation of any engine. Since the atmosphere is almost one-quarter oxygen by mass, running the advanced engines in the atmosphere is a convenient way to avoid having to carry that oxygen along inside the ship. From the standpoint of engine operation alone, the spaceplane should be able to stay in the atmosphere a long time.

The next item on the FFM list is *working mass*, which the atmosphere also has in unlimited abundance. This working mass is the sum total of the atmosphere, composed mainly of inert nitrogen, but also trace elements, in addition to the oxygen that we have already considered. What is so important about this working mass? Well, it serves as the propellant, the stuff that propels the machine forward, as long as the flying machine remains in the atmosphere. Like the oxygen for the engine, the working mass does not have to be carried onboard as long as the vehicle is flying through it.

The turbine in a typical jet engine powers a compressor that acts like a huge vacuum cleaner, sucking in air, compressing it, and eventually shooting it out the rear as a jet of continuous thrust. The turbine uses fuel (typically a kerosene derivative) and atmospheric oxygen while the compressor utilizes the working fluid of the surrounding air mass. Again, from the standpoint of working mass, it is expedient for the spaceplane to remain in the atmosphere as long as desired.

At some point, the advanced spaceplane must accelerate to orbital speed. But increasing speed also increases drag, the arch-nemesis of all aircraft. Supersonic speeds also result in severe heating, another serious barrier to overcome. At some point, therefore, the spaceplane must exit the atmosphere to remain flight efficient, and this will probably occur long before reaching orbital speed.

An optimum velocity profile during ascent to orbit involves deciding how fast to fly at every altitude, and at what angle to climb, in order to minimize onboard propellant usage. Subsonic cruise-climb is used during the first part of the flight, because this achieves about 10% of orbit insertion altitude and minimizes drag both during this stage and the later zoom to orbit. This ascent profile is similar to that of a modern airliner or strategic bomber, and takes the spaceplane from the runway up to some 50,000 ft. At this point, still climbing, the spaceplane applies full thrust, accelerates to supersonic speeds, and pitches up to around a 40-degree angle of attack. The air-breathing turborocket remains effective until speeds reach the low hypersonic region – Mach 6 or 7. The pilot now switches off the helium loop; the compressor shuts down, and the spaceplane goes to pure rocket thrust. By now, it has gone completely ballistic, is leaving the sensible atmosphere, and has only its onboard propellants to reach orbital speed. The spaceplane pitches over to more nearly horizontal, settles into the proper angle of attack to counter remaining gravity losses, and aims for its orbital insertion point downrange.

By using subsonic lift up to 50,000 ft, and then employing a 40-degree angle of climb thereafter, gravity losses are kept to a minimum, because aerodynamic lift is used to the maximum practical extent. At the same time, atmospheric drag and thermodynamic loads are kept at bay by climbing subsonically in the lower atmosphere, and by using a greater rate of climb during the supersonic zoom to orbit. In other words, by using this optimum velocity profile, both gravity and drag losses can be kept low, while making full use of the atmosphere for lift, oxygen, and working mass.

Another aspect of FFM involves refueling, which actually refers to aerial propellant transfer or aerial propellant tanking (APT). By pulling up behind a tanker aircraft, a spaceplane can take on its load of oxidizer-propellant at altitude rather than taking off from a runway fully loaded (Fig. 9.17). An oxidizer is typically the heaviest item onboard any space launch vehicle, and so departing the ground without it would have obvious benefits. The oxidizer to be transferred could be liquid oxygen (LO_2) or it could be hydrogen peroxide (H_2O_2), which looks and handles much like water, although it is highly sensitive to impurities. Alternatively, the spaceplane might generate onboard oxygen during subsonic flight by liquefying ambient air, separating out the oxygen component, and storing it internally. This last method involves much longer subsonic flight times and requires onboard lightweight air-liquefaction equipment. All of these are possible technologies for the advanced spaceplane.

Lightweight composite materials will make up much of the advanced spaceplane's structure. It is vitally important that the design be both lightweight and very strong, in order to achieve the necessary mass ratio for single-stage-to-orbit flight. One element of the design will combine a high strength aeroshell with the thermal protection system. By using a double-walled honeycomb structure, the spaceplanes

Fig. 9.17 Spaceplanes may refuel in flight after taking off, like this B-2 Bomber being refueled by a US Air Force KC-135 tanker (courtesy USAF)

of the future will have highly reliable thermal protection systems and strong reentry bodies, both vital for Earth orbital and Lunar return reentries. This design was first considered for use on the X-33 testbed, before its ignominious cancellation.

For wingless rockets, the mass ratio required for SSTO spaceflight is about 8.5, using high-energy propellants and altitude-compensated rocket engines. This means that no more than 12% of the vehicle's gross lift-off weight can be composed of structure, engines, crew, and cargo. The other 88% must be propellants only. Applying the rocket equation, and assuming a specific impulse of 435 s, we can calculate the following result.

$$\Delta V = c \ln R \text{ where } c = I_{sp} g_e \text{ so}$$
$$\Delta V = I_{sp} g_e \ln R$$
$$\Delta V = (435 \text{ s})\ (32.174 \text{ ft/s}^2) \ln (8.5)$$
$$\Delta V = 29{,}950 \text{ ft/s}$$

This result is equal to 5.67 miles/s, or 9.13 km/s, significantly higher than low Earth orbital velocity. When drag and gravity losses are taken into account, we find that this is about right. This calculation is based on pure rocketry, with no FFM, lifting optimum velocity profile, or APT involved. When all the tricks of the trade of the advanced spaceplane are used, it will be possible to reduce this mass ratio significantly, thereby increasing the payload fraction to practical levels. We saw an example of this in the last chapter.

Advanced spaceplane infrastructure will involve ground infrastructure in the form of spaceports, aerial infrastructure in the form of airborne tankers, and space infrastructure in the form of orbiting propellant depots, space stations, and Lunar bases. This infrastructure is already well on the road to developing, since existing airports will become future spaceports. Aerial tankers already exist as well, and in-flight refueling procedures are well-practiced and routine. With the establishment of the first low Earth orbital "gas stations," together with the first spaceplane "customers," spaceflight activity will rapidly increase.

Getting There

How do we reach this vision of the future? How do we get there in one piece? The key that will open up the door to this kind of future is the reusable advanced spaceplane. Extensive flight test, leading to increased reliability and safety, advanced engines, superior structures, and competent operations all lead to a satisfied market, which will pay for the entire process.

Good spaceships (Fig. 9.18) are a prerequisite for safe, reliable space transportation. The advanced spaceplane will be that spaceship, without relying on the primitive, wasteful concepts of rocket staging or modular missions. They will be the final stage in an evolution from primitive rocketplane, through suborbital spaceplane, to

Fig. 9.18 Is this what good spaceships of the future will look like? This is the BOR-4 unmanned Russian spaceplane on display (courtesy http://www.buran-energia.com)

single-stage-to-orbit spaceplane. Only by flying thousands – not hundreds – of missions, only by proper FFM techniques, only by using superior structures and engines, and only by cool competence and sustained development will this come about. The suborbital spaceflight experience gained by the operator, as well as enjoyed by the space tourist, is the key that will open the spacelanes and stimulate the spacelines of the future.

Chapter 10
Destination Moon: The Lunar Spaceplane

Moonships of the future will be spaceplanes. It is easier for spaceplanes to fly to the Moon than it is for spaceplanes to get into Earth orbit itself. It is even easier to fly back to Earth. Just design a single vehicle, fly it to the Moon, and fly it back – simple, operationally sound, reliable. But can it actually be done?

Is it even *rational* to attempt Lunar spaceflight with a winged space vehicle? After all, the added weight of wings, vertical stabilizer, landing gear, and aircraft-like structure all amount to "dead weight" that must be lifted – and *accelerated* – to the Moon. These arguments are undeniably sound, but they skirt the real issue, which is reusability. And, as we have seen, reusability is the key to spaceflight operations. For a space vehicle returning from the Moon to be fully reusable, it must have wings. Blunt reentry bodies such as those used in Apollo and now being developed for Orion are inadequate, because reusability, if it exists at all, is only partial. Furthermore, partial reusability requires operational complexity, and operational complexity is the bane of future space access. With this in mind, how might an operationally simple Lunar spaceplane work?

Lunar Vision

Once an advanced spaceplane has reached low Earth orbit (LEO), its orbital sojourn is normally limited only by the stores of food, water, oxygen, and maneuvering propellants onboard. Its main propellants have been exhausted, making it essentially an empty winged bulk tank, orbiting just outside the atmosphere. In this condition, the spaceplane's only option, at some point, is to reenter Earth's atmosphere and land.

But consider how dramatically the picture changes when the spaceplane is refueled *in orbit* instead of on the ground. With an existing kinetic advantage of 5 miles/s (not to mention a potential energy store of some 200 miles altitude), an orbitally refueled spaceplane will have an immediate capability of undertaking any space mission requiring a delta-V of five *additional* miles per second. Yet Lunar missions (and indeed, all Earth escape missions) require only a fraction of this amount – about 2 miles per second. This opens up huge potentialities: Breaking out

M.A. Bentley, *Spaceplanes: From Airport to Spaceport*,
doi:10.1007/978-0-387-76510-5_10, © Springer Science+Business Media, LLC 2009

of Earth orbit requires only two-fifths the delta-V that it took to get into orbit in the first place. As a result, an orbitally refueled spaceplane would have far more than enough propellant to fly all the way to the Moon and back. The extra propellants can be used to replenish Lunar landing craft, increase the space maneuvering capability of the spaceplane, and even reduce the atmospheric reentry speed on the return flight. The remaining reserves can be used for maneuvering inside the atmosphere, including making missed approaches or go-arounds during the landing phase. All of this adds up to increased safety.

Future spaceplanes will doubtless have the ability to land directly on the Moon, but initially some form of Lunar orbit rendezvous will likely be used. Lunar landers will shuttle back and forth between the Moon's surface and Lunar orbit. Spaceplanes, with their enormous propellant tanks, will refuel the landers in orbit, transfer Moon-bound passengers, and take on Earth-bound ones. Not only will they transport eager space tourists to and from Lunar orbit, but they will also serve as supply vessels for Lunar infrastructure. In this way, spaceline companies will optimize the economics of their operations.

A typical Lunar spaceplane mission will take about 1 week. Let us take a look at how such a mission might unfold. The 50-passenger spaceplane is pulled away from the gate and towed to the end of the runway by ground-based vehicles. This minimizes waste of precious propellant prior to takeoff. Once cleared for departure, the pilots start the air-breathing turborockets, advance the throttles, and release the brakes. The plane quickly accelerates and rapidly thunders skyward, using the atmosphere at this point for most of its propulsion. Within 10 min it has pulled up behind an aerial tanker carrying the heavy oxidizer propellant that the spaceplane will need to make orbit. Aerial propellant transfer of liquid oxygen (LOX) or hydrogen peroxide (H_2O_2) takes another 10 min, followed immediately by all-engine light-off. The vehicle pitches up to the optimum flight attitude, thereby minimizing gravity and drag losses while maximizing the benefit of aerodynamic lift. In moments, the vehicle has exited the sensible atmosphere and is operating as a pure space vessel. The aerospike engines provide altitude compensation from the upper atmosphere to orbital altitude, squeezing every ounce of efficiency out of the propellants. This, together with the earlier air-breathing portion of the ascent, provides for the critical increase in specific impulse and delta-V required to reach orbit. Within 30 min of brake release, the spaceplane is cruising at an altitude of 100 miles and a speed of 17,500 mph. It has reached LEO. Some 45 min later, upon reaching the 200-mile apogee, the captain circularizes the orbit with another short burn of the engines.

The next phase of the flight involves orbital propellant transfer (OPT), which means rendezvousing with an orbital fuel depot. Regular flights of spaceplane tankers keep the depot supplied on a continuous basis. Station-keeping with the orbital cache, a space robot plugs fuel and oxidizer lines into refueling ports in the spaceplane. A third line transfers pure water. Solar-charged batteries drive pumps to transfer the propellants, and soon the vehicle is ready for trans-Lunar insertion.

For most of the passengers, who are leaving Earth for the first time in their lives, the thrill of breaking out of Earth's gravitational grip and seeing the home planet floating in space is awe-inspiring. The sight of Earth as a beautiful oasis in space, as experienced by millions of future space tourists, will surely change forever humankind's perspective of themselves and their place in the universe. The benefits from these kinds of experiences, repeated many times over for people from around the globe, can only result in a better future for the entire human race.

Now the fun begins. Passengers look forward to 3 days of free-fall on the uphill coast to the Moon, during which they are free to float and cavort about the cabin, gaze through the many portholes at Earth, the Moon, or the stars, and generally accustom themselves to the weightless environment. Every spaceliner has an observation deck equipped with telescopes for the viewing pleasure of all passengers. Through these, everyone enjoys unobstructed views of the heavens.

As the ship nears Luna, cabin chatter is abuzz in anticipation of seeing the far side of the Moon, which is never visible from Earth. All eyes are glued to the portholes as the ship arcs around the Lunar limb and Earth disappears from view. For the first time in their lives, passengers cannot see their world. One of the most conspicuous features of the far side is the huge crater *Tsiolkovskiy*, named after the famous Russian rocket scientist (Fig. 10.1). Presently, the captain announces that all passengers and crew are to take their seats and buckle in securely for the deceleration maneuver into Lunar orbit. The rocket burn is accomplished by turning the spaceplane around so that it is flying backward, then firing the rearward-facing maneuvering engines. To the passengers, it seems as if they are accelerating forward rather than decelerating backward. The flight crew checks their instruments to ensure that

Fig. 10.1 Tsiolkovskiy Crater on the Lunar far side, as photographed by the crew of *Apollo 8*. This feature was discovered by the unmanned Russian space probe Luna III in 1959 and named in honor of Konstantin Eduardovich Tsiolkovskiy, father of Russian cosmonautics. The feature measures some 250 miles in diameter (courtesy NASA)

the proper orbit has been achieved, and passengers are once again released from the confines of their seats. After seeing Earth-set and the far side of the Moon, the next event is Earth-rise above the desolate Lunar hills below. Cameras click away as each space tourist vies for his or her own version of the famous Apollo 8 Earth-rise picture, taken by the first Earthlings to fly to the Moon.

After circling the Moon perhaps one more time, passengers are once again instructed to take their places and secure their safety belts. It is time for rendezvous and docking with the Lunar shuttle. As passengers crane their necks for a better view, some catch a glimpse of an ungainly vehicle completely lacking any semblance of airworthiness. This is the vessel that will transport them from the sleek spaceplane to the dusty surface of the Moon. As this vehicle draws nearer and nearer, its strange outlines become apparent. It looks more like a crab than anything else, with a short and squat body, six gangly legs, and various antennae sprouting from its arthropod-like midsection. Yet the design is eminently spaceworthy and is intended to provide the greatest level of safety, comfort, and efficiency for operators and passengers alike. Its overall dimensions – short and squat – ensure that it can never tip over, even if it should land on the slope of a crater. Like the spaceplane, it carries 50 passengers and a crew of five. Upon docking, two hatches are opened between the vessels, so that passengers can be transferred quickly from vessel to vessel with a minimum of confusion. A total of 100 passengers are efficiently "transfloated" in a well-practiced sort of space ballet. At the same time as this is taking place, pure water is transferred from the spaceplane to the thirsty Lunar crab. This will be refined into rocket propellants after landing, to be used by this and other Lunar shuttles.

This vision of the near future is a glimpse of a relatively immature Lunar infrastructure. Propellants are being brought from Earth to the Moon, rather than from the Moon to LEO, and Lunar spaceplanes are not yet sufficiently developed to land on the Moon themselves. Nevertheless, Lunar tourism is in full swing, with 50-passenger Lunar shuttles making regular flights to and from Lunar orbit.

Why Go to the Moon?

At the first NASA Lunar settlement conference in 1984, the late German-American space scientist Dr. Krafft Ehricke proclaimed, "If God wanted man to become a spacefaring species, he would have given man a moon."[1] Well, there it is (Fig. 10.2). The Moon is the perfect place to practice the science of spaceflight and the art of space living. It is just far enough away (30 Earth diameters) to make the crossing a challenge. Its gravity is weak enough (1/6 of Earth normal) to make landings and liftoffs relatively easy. It is perched high enough on the rim of Earth's gravitational well, that returning home is also easy, but spacecraft are still challenged by the problems of deceleration and atmospheric reentry. The Moon is also the perfect staging ground for the interplanetary flights of the future. And its environment is just harsh enough that, if you are incompetent, you will perish. Yet it is still close enough that rescue ships can arrive within days. The Moon is the perfect

Fig. 10.2 The Moon has always inspired the romantic, and beckoned the bold. This is what future travelers to the Moon might see just before the Trans-Lunar Insertion burn (courtesy NASA)

real-life laboratory to practice flying in space and flying between planets. It cannot be denied that we humans are indeed fortunate to have a large Moon roughly the same size as the very largest moons in the Solar System.

The main reason for returning to the Moon, and going there in good spaceships, is because *it is there*. Not even Christopher Columbus knew there was an entire New World awaiting his discovery, sitting in the middle of the world ocean between Europe and the East Indies. But he still sailed fearlessly into the pages of history. He could not have made this discovery without good ships. Nearly 500 years before the voyages of Columbus, another group of seafarers – the Vikings – discovered the same landmass, hopping their way from Norway and Ireland, to Iceland, then to Greenland, and finally to the coast of North America. They also had good ships. The Vikings and Columbus had something else in common: the skills to sail their ships proficiently, and the desire to boldly venture into the unknown. Again, the main reason they went was because the place *was there*, and they were curious. Neither the Vikings nor Columbus knew the potential of the new continent. And we do not yet know the full potential of the Moon.

Missiles and modules can get us to the Moon. But what are really needed are good spaceships that are not designed to come apart or be thrown away. The advanced spaceplane described in the last chapter is just such a spaceship. With the proper utilization of resources, good spaceships will open up the Lunar and interplanetary frontier without wasting hardware or turning the Lunar landscape into a government dumping ground. Let us explore what the Moon has to offer, how we got there the first time, and what is in store as we consider venturing back.

Lunar Resources

Earth has no business with a moon as large as Luna. Our Moon, in fact, is so large compared to its parent planet that some have considered calling the pair a double planet. Spaceflight between Earth and Moon is similar to interplanetary flight on a greatly reduced scale. The Moon is larger than Pluto, and almost as large as Mercury. It is currently some 240,000 miles away from Earth, but this was not always the case. It is receding at the rate of about 1 in. per year, so that by "running the clock" backward, scientists have deduced that the Moon originated much closer in. Some have even suggested that the Moon was born of our home world, mysteriously ejecting itself out of what is now the Pacific Basin long before the reign of the dinosaurs or the trilobites, or even the first unicellular organisms. They point to the volcanic "rim of fire" that still encircles the Pacific as proof of this thesis. But this is an outmoded idea, borne of the simple-minded notion that current trends – the Moon's incremental retreat – can be time-reversed and extrapolated back to a planetary fission. Using sophisticated computer-generated trajectory and collision analyses, the latest theory of the Moon's origin is that the proto-Earth was struck by a pre-Lunar body roughly the size of Mars early in the formation of the Solar System. This event is supposed to have occurred about 4½ billion years ago. The collision of the two planets caused a coagulation and redistribution of mass on a planet-rending scale, completely obliterating the smaller body, while simultaneously throwing into orbit a huge mass of debris that initially became a ring system. The young Earth, fuming molten red from this unexpected turn of events, would have looked like Saturn's little sister. Shortly – in a mere million years or so – the planetary ring system coalesced and became the Moon. This theory explains why the Moon does not orbit in the plane of Earth's equator, and might even help to explain why the Lunar core is so small. The Moon is composed mainly of mantle-like material ejected from the proto-Earth.

What does the Moon have to offer us? As mentioned above, the greatest gift is the potential the Moon has in helping mankind become a spacefaring species. The Moon, for a planet like Earth, provides the perfect base of operations for the future development of a space-based civilization. It promises to give us the sheer experience we will need to become masters of the Solar System, in the same way that Scandinavian geography made the Vikings masters of their northern realm. This is an intangible resource, but a physical one nonetheless, for the Moon is very real and very near. Beyond the practice in spaceflight operations we will gain in going to the Moon, there is also the nitty-gritty stuff that the Moon is actually made of. There are elements in the regolith that do not exist on Earth. There may be large quantities of ice lurking in the polar shadows. There could well be undiscovered caves harboring unsuspected treasures. Imagine finding an alien artifact, a crashed spaceship from some planet beyond the Solar System, or even an alien corpse. For the scientist, the Moon is a natural laboratory. For the geologist, it is a paradise. For the astronomer, the views of the heavens are unhampered. For the radio astronomer, the Lunar far side offers a "quiet" environment free of artificial Earth emissions. And for the biologist, it offers a harsh environment to test the survivability of various organisms. The botanist will practice perfecting plants that can survive – even thrive – through the 2-week-long Lunar night. The rocket scientist will synthesize energetic propellants directly from the Lunar soil, or electrolyze them from local ice deposits. Environmental engineers, too, will derive life-sustaining oxygen from the same sources. And the nuclear engineer will mine the Moon for the rare-Earth element helium-3, to be used as fuel in fusion reactors on the home world. These are just a few of the resources our good satellite has to offer, but we will need good spaceships to get us back and forth.

To sustain spaceflight operations between Moon and Earth, a Lunar source of rocket propellants would be a great boon. There are two obvious sources worth investigating: the polar regions and the ubiquitous regolith, or Lunar soil. The Clementine spacecraft detected evidence for water in 1995 at the Lunar poles. This could be interpreted as evidence for ice deposits in shaded areas where the Sun never shines. And Apollo astronauts brought back samples that reveal the abundance of helium-3, a heavy isotope of helium.[2,3]

Water, in the form of ice, is abundant in space. This may seem surprising, but it is borne out by the fact that the Solar System is full of icy bodies. The Kuiper Belt, that region inhabited by Pluto and other dim planetoids, might just as well be called the Ice Belt. Further out is the Oort Cloud, the vast repository of the comets. And these are icy bodies as well. Three of the large moons of Jupiter – Europa, Ganymede, and Callisto – have large quantities of ice. Europa is believed to have a liquid water ocean vaster by volume than the seven seas of our own world, permanently hidden beneath an ice crust. And Mars has huge quantities of ice at its poles and beneath its surface. All of this supports the expectation that water ice will be found on Luna as well. NASA's Lunar flight program is concentrating on the polar regions in hopes of finding ice deposits in permanently shaded craters. If such deposits do exist, then it will be possible for astronauts to "live off the land" by mining the Moon for ice. It can then be melted, purified, and used for a host of purposes. Among these are drinking water, rocket propellants, and oxygen for life support.

For 4½ billion years, the Moon has been absorbing particles from the Solar wind and imbedding them in the dusty regolith. Among these particles are atoms of helium-3, a specific isotope of that inert element first discovered in the spectrum of the Sun itself. Helium-3 occurs on Earth in tiny quantities only, but it exists in much greater abundance on the Moon. This isotope – or heavy version – of helium may one day power a nuclear fusion technology for the twenty-first century on Earth. And such a civilization would depend on the Moon for its supply of helium-3.

Mining the Lunar regolith for helium-3 could have unexpected benefits that pertain directly to space infrastructure. As this rare-Earth element is refined, several important by-products appear during the process. These include large amounts of hydrogen, as well as methane, carbon dioxide, and water. These compounds, or the elements they contain, are locked in the Lunar soil, and have obvious uses as propellants and life-sustaining substances.[4]

Project Apollo

The fact that human beings reached the Moon and walked on its pristine surface just 8 years after President John F. Kennedy committed America to that goal is nothing short of amazing. The first manned spacecraft to reach the Moon was Apollo 8, at Christmastime 1968. The crew was Frank Borman, James Lovell, and William Anders. As they orbited the Moon that Christmas Eve, they each read passages from the first chapter of Genesis, a fitting gesture as they flew in heavenly proximity to the handiwork of creation. I remember that flight, going outside to see if my 7-year-old eyes could spot the spacecraft near the Moon. I was sure I saw it – a tiny speck, an insignificant dot very close to our good satellite. Such is the expectant vision of youth (Fig. 10.3)!

Fig. 10.3 Apollo 17 Saturn V rocket poised for liftoff to the Moon, December 1972. This was the last Lunar flight of Apollo (courtesy NASA)

This first flight to the Moon was a bold move, sending three brave men hurtling away from Earth like never before. The Apollo 8 spacecraft flew without a Lunar Module, because the subcontractor, Grumman, was still perfecting the final details. This meant that if anything went wrong with the basic spacecraft – the combined Command and Service Modules – the crew could well be stranded in Lunar orbit, or worse yet, in Solar orbit. But everything went as planned, and the crew returned safely.

For Apollo, missiles and modules were the key to success. This, together with a brand new technique called Lunar orbit rendezvous, enabled the first manned Moon shots before the year 1970. For Lunar landing missions, there were eight separate rocketships that helped man reach the Moon. These were the three main launch stages, the launch escape tower, the Moon ship, and the Command ship. The last two were made of two pieces each. By leaving the descent stage of the Lunar Module on the Moon, and rendezvousing with the Command ship in the ascent stage, a minimum of weight would have to be rocketed off the surface of the Moon. By minimizing the various landing or liftoff weights, the total amount of required propellant could be kept down as well, making the job that much easier for the Saturn V launch vehicle. The Saturn V was the largest rocket built to date. It stood 363-ft tall, weighed 6½ million pounds at liftoff, and was able to generate 7½ million pounds of thrust. It rose slowly, majestically off the launchpad, but afterward accelerated rapidly to boost its payload into orbit.

Why did the missiles and modules concept work so well in getting Apollo to the Moon? The entire thrust behind this idea was to launch each mission as a self-contained unit and to minimize propellant usage during the flight. There was never a plan to perform in-space refueling. The Saturn V lifted off with everything the mission would need, every drop of propellant, every piece of equipment that it would take to get two astronauts onto and off of the Moon while a third waited in Lunar orbit. The plan worked admirably, but there was a price to pay. Not one piece of the Apollo/Saturn space vehicle was reusable. Every piece was thrown away after it had done its job. So, although the astronauts always got to fly brand new equipment, the costs were tremendous.

It took three enormous rocket stages to get the Saturn V stack off the ground. The first stage used five huge F-1 engines burning kerosene and LOX and delivering 1½ million pounds of thrust each. The dense fuel kerosene was used here because it could deliver greater thrust to boost the vehicle off the pad, although the propellant consumption in terms of mass flow rate was higher, and the specific impulse was lower than with liquid hydrogen. Still, at this stage of the flight raw thrust was more important than specific impulse, because the launch vehicle was still moving relatively slowly. When the first stage burned out and dropped off, the five J-2 engines of the second stage took over, burning LH_2 and LOX, now that the speeds had increased to the point that specific impulse became more important than thrust alone. Remember that specific impulse is simply a rocket's thrust divided by its propellant weight flow rate. Liquid hydrogen, with its higher specific energy content and lower density compared to kerosene, delivered a higher specific impulse. So this was the fuel of choice for both the second and third stages. Once the second stage had expended all of its propellants, the lone J-2 engine of the

Saturn's third stage finished the job and delivered the remaining stack into LEO. This engine could be shut down and restarted, a feature that allowed the third stage to boost the Apollo spacecraft complex to the Moon.

Once the third stage, the S-IVB, had done its job and the spacecraft was established on its high-speed coast toward the Moon, the Command ship disengaged, turned around, and docked nose-to-nose with the Lunar Module. The Command ship then pulled the LM out of its docking adapter – the shroud that had protected it during the atmospheric ascent – and together they continued their uphill ride. The S-IVB was now cast aside, since it could serve no more useful purpose. See Chap. 4, Figs. 4.7 and 4.9.

At this point, the original eight-piece stack was down to four. The launch escape tower – intended to pull the crew capsule away from the stack in case the rocket blew up – had been jettisoned during the boost to orbit (Fig. 4.8). The remaining modules consisted of the Command Module, the Service Module, and the Lunar Module descent and ascent stages. The service propulsion system engine, with its large area-ratio nozzle mounted at the aft end of the SM, would do the job of getting the spacecraft into and out of Lunar orbit. If this engine failed, especially if it failed after the Lunar Module's propellants had been expended, the crew would be marooned in space. Once in Lunar orbit, two of the astronauts transferred from the Command ship into the Lunar lander, undocked, and flew down to the surface. Using the descent stage's engine only, they navigated to a preselected touchdown zone. Coming out of orbit, the LM used its descent engine initially to reduce orbital, or forward, speed (Fig. 10.4). This allowed the Moon's gravity to pull the craft toward the surface. By carefully adjusting the attitude of the lander, the descent engine could be used to reduce vertical velocity as well as horizontal, so that just before touchdown, the LM was hovering with little or no forward speed, over the landing zone (Fig. 10.5).

When Lunar explorations were complete, the crew climbed back into the ascent stage, used the descent stage as a launchpad, and blasted back into orbit. If this engine failed for any reason, the crew would be stranded on the Moon. The two Moon explorers now met up with the lone astronaut in the Command ship (even though the commander of the mission got to walk on the Moon), transferred rocks and samples, and jettisoned the LM ascent stage.

Now the spacecraft, with all three astronauts back together, was down to two modules. The Service Module's high area-ratio engine (tailored for efficiency in vacuum conditions) pushed the spacecraft out of Lunar orbit to begin the long downhill coast back to Earth. By the time the spaceship reached the atmosphere, it had accelerated to around 25,000 mph, or 7 miles per second. The cylindrical Service Module was discarded just before reentry, with the conical Command Module entering the atmosphere alone, protected by an ablative heat shield and a detached bow shock wave. The astronauts endured 6.8 *G*'s during this phase of the mission, until drogue and main parachutes were deployed and the Apollo crew capsule splashed down at sea.

Unlike the pilots of the Space Shuttle, or any aircraft for that matter, the crew now relied completely on rescue–recovery from the US Navy (Fig. 10.6). Frogmen

Fig. 10.4 Multiple exposure photograph showing how the Lunar Module approached its touchdown zone, hovering briefly over the landing point before settling to the surface (courtesy NASA)

were dispatched from helicopters, flotation collars were attached to the spacecraft, and life rafts were inflated for the astronauts. The frogmen helped the crew out of their spacecraft, into the life raft, and finally onto the deck of an aircraft carrier.

The missiles and modules of Apollo won the space race to the Moon. The system worked beautifully, and Apollo accomplished its objectives. For its time, the modular concept was the correct choice. But what of the future?

Project Constellation

The United States plans to return to the Moon within 12 years, as part of NASA's Project Constellation. A new series of launch vehicles and spacecraft are being developed to get humans back to the Moon by the year 2020. The launch vehicle, Ares, will use both solid- and liquid-fueled rocket stages. The spacecraft Orion will

Fig. 10.5 *Apollo 16* astronaut John Young launches himself off the ground in the weak Lunar gravity as he salutes the Stars and Stripes. The Lunar Rover can be seen parked by the Lunar Module (courtesy NASA)

Fig. 10.6 *Apollo 11* after splashdown, awaiting recovery by the US Navy. This was the mission that landed the first men on the Moon (courtesy NASA)

be an enlarged version of the Apollo Command Module, with a crew capacity of up to six for Earth orbital flights, or four to the Moon. Similar to the Soyuz spacecraft, the Orion Service Module will include deployable solar panels. Similar to the Space Shuttle, the Ares boosters will be reusable solid-propellant rockets, extended to five segments. Similar to Apollo, the upper stages will use liquid propellants powered by the new J-2X engine, an outgrowth of the old Apollo J-2. Unlike Apollo, each Lunar mission will involve two launches and a conical Crew Module that can be reused up to ten times. Although original plans called for a parachute landing on dry land, the plan now is to splash down in the eastern Pacific, much like Apollo.

As of this writing, the Lunar Return Flight Plan looks like this. An unmanned Ares V rocket (Fig. 10.7) lifts off from Cape Canaveral with a fully fueled Earth Departure Stage (EDS) and a Lunar lander called a Lunar Surface Access Module (LSAM). The LSAM is made of two modules, just like the Apollo LM. These modules are placed in LEO to await the arrival of the crew. Shortly after, an Ares I rocket with an Orion spacecraft and crew of four lifts off and rendezvous' with the

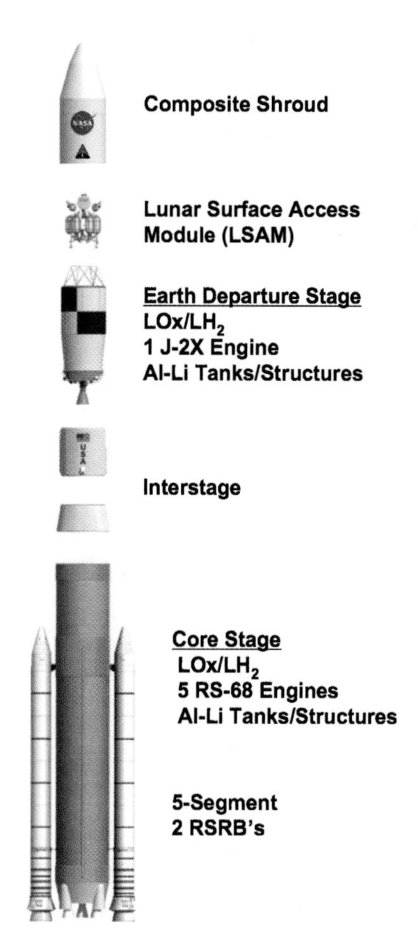

Fig. 10.7 Ares V unmanned rocket, used to launch heavy loads, including Lunar transfer stages and Moon landers. The smaller Ares I (Fig. 2.7) will launch astronaut crews (courtesy NASA)

Fig. 10.8 Earth Departure Stage (EDS) and Lunar Surface Access Module (LSAM) in Earth orbit, with Orion Crew Module attached, ready for trans-Lunar insertion (courtesy NASA)

unmanned vehicle. Orion docks with the EDS and LSAM (Fig. 10.8), upon which the combined spacecraft rocket to the Moon using the EDS engine. Similar to the S-IVB, the EDS is cast adrift after expending all its propellants. Similar to Apollo, there are now four modules coasting uphill to Luna. Unlike Apollo, docking with the Lunar lander takes place in LEO before the EDS burn rather than on the way to the Moon after the S-IVB burn. Upon reaching the far side of the Moon, the LSAM's engines, instead of the Service Module engine, are used to brake the docked spacecraft into orbit. Next, all four astronauts transfer into the LSAM and descend to the Lunar surface for about a week of exploration and surface activity, leaving the empty Orion in Lunar orbit. In Apollo, the Command Module pilot remained in Lunar orbit while the other two astronauts explored the surface. The LSAM will be an enlarged version of the Apollo Lunar Module, with a four-legged descent stage and a manned ascent stage. Instead of a single engine in the descent stage, the LSAM will have four engines, as well as a single engine of the same type in the ascent stage. Similar to the LM, the LSAM is to be used one time only. The landing sites will be near the poles, in hopes of finding water ice in those regions. Similar to Apollo, there is a conspicuous lack of in-space refueling.

What is original about the Ares/Orion Lunar landing plan? Well, the Crew and Service Modules are virtual duplicates of the Apollo CSM, enlarged and upgraded

Fig. 10.9 The Lunar Surface Access Module showing the ascent stage lifting off after a surface expedition. The LSAM will carry the entire crew of four to the surface (courtesy NASA)

with better computers, fan-shaped solar panels, and reusable up to ten times. The LSAM (Fig. 10.9) is a beefed-up version of the Apollo LM, with the addition of three more rocket engines in the descent stage, but not refuelable or reusable. The EDS is the equivalent of an enlarged S-IVB, even using the same J-2 engine upgraded to J-2X. The launch escape tower is Apollo-era technology, and the SRBs are borrowed from the Space Shuttle, lengthened by one segment. The answer is nothing. Not even the procedures are new. Nothing is being done for the first time. The Apollo program, including the precursor Mercury and Gemini projects, invented almost everything: man in space (women in space had to wait for the Space Shuttle, Valentina Tereshkova notwithstanding), extravehicular activity or "spacewalks," orbital rendezvous and docking, ablative heat shields, splashdown and recovery operations, and landings on the Moon. Project Constellation presents nothing new, not even in-space refueling. This means that as each module runs out of propellant, it becomes a useless hollow hulk, a space derelict. Project Constellation will, however, almost certainly take the first women to the Moon, which is noteworthy.

The only components of Project Constellation that are planned to be reusable are the solid rocket boosters and the Orion Crew Module (Figs. 10.10 and 10.11). Compared to Apollo, this is laudable. The Ares/Orion spaceflight concept will be much safer and simpler than the Shuttle. If anything unfortunate happens during liftoff, the launch abort system (Apollo had a launch *escape* system, same thing) will activate and accelerate the crew capsule safely away from a malfunctioning booster. Also, with Orion perched at the top of the stack, the danger from falling debris has been eliminated. But it is still missiles and modules.

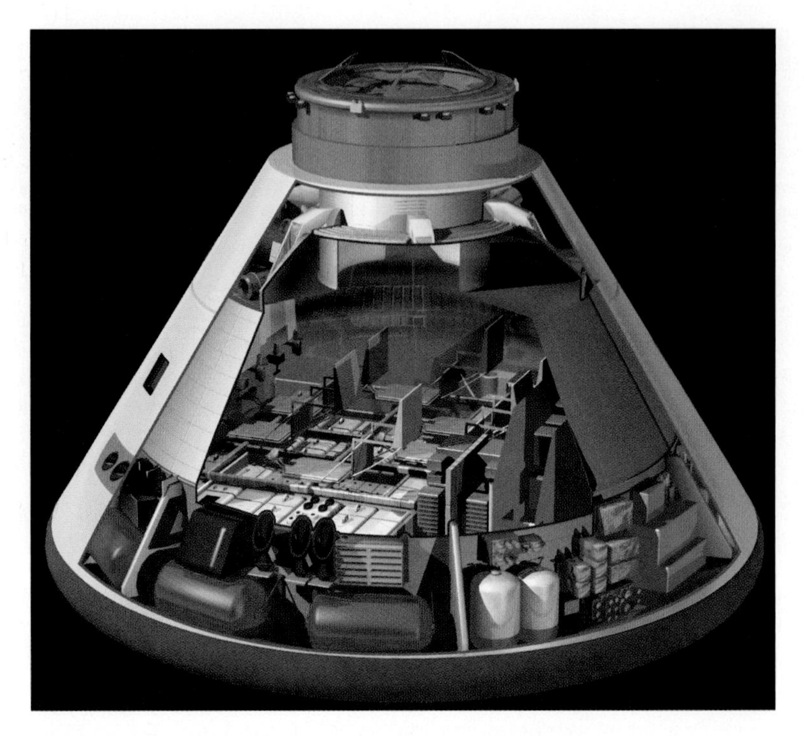

Fig. 10.10 The Orion Crew Module, an enlarged and improved version of the Apollo Command Module, is to be used for flights to the International Space Station, the Moon, and even to Mars (courtesy NASA)

Fig. 10.11 The Orion Crew Capsule enters Earth's atmosphere prior to splashdown at sea or "bump-down" on dry land. Either way, large parachutes are required because the return module has no wings (courtesy NASA)

Good Spaceships

To sustain a thriving Lunar transportation system in the future, we will need good spaceships. And a good spaceship, by definition, is one that is fully reusable. As we have seen, the best way to make an Earth orbital spaceship reusable is to give it wings. It has to be a spaceplane. Also, as we have seen, the SSTO spaceplane has large propellant tanks, making it the ideal space tanker. The conclusion is that among the best spaceships of the future will be winged space tankers. This automatically makes them a part of the infrastructure needed to refuel *other* good spaceships. As for good spaceships operating near the Moon, the situation is different. The Moon has no atmosphere, so good Lunar spaceships do not need wings, as long as they stay there. Also, Lunar gravity is only one sixth that at the surface of Earth, and so good spaceships need far less fuel, but to be good spaceships, they have to *refuel*. This means that some other ship has to bring the fuel, unless fuel is found on the Moon. If propellants are found on the Moon, then they will have to be transported, in some form, to LEO, as we will see. What does all this have to do with the Lunar spaceplane? The answer is obvious. *Spaceplanes are uniquely qualified to ship propellants both to and from the Moon.*

For any good spaceship to operate between Earth and the Moon, it must be able to use the Earth's atmosphere for aerobraking. This means that all good spaceships returning from the Moon must be designed aerodynamically. Using blunt body capsules is one solution, but an inefficient one, because they are modular and therefore incomplete. The spaceplane, whether it sports a double-delta wing, a lifting body, or a blended wing-body, is far superior to any kind of reentry capsule. Spaceplanes are simple, complete, self-contained, in-space refuelable, fully reusable, and versatile. They are good spaceships, rather than piece-meal spacecraft.

A comparison of the modular Moon mission to the Lunar spaceplane reveals some interesting facts in terms of mission complexity. As every engineer knows, complexity often leads to failure; so the simpler a system is made, the more reliable it will be. And yet, the modular concept is complex, because it involves so many components. Furthermore, the modular mission is inefficient, because it repeats the same things over and over. For example, modules first use one set of propellant tanks and one engine, then (after they throw those away) they use another completely different set of propellant tanks and another engine, each with its complex pressurization system, piping, valves, nozzles, gimbals, etc. Of course, the reason they do this is because they do not refuel in space. Ironically, by using multiple modules to keep mass and propellant usage down, total mass is actually driven up. Table 10.1 shows a simple comparison of three Lunar spacecraft, the Apollo, Orion, and spaceplane, as they prepare to leave Earth orbit. The spaceplane, good spaceship that it is, has just refueled, and so it is equipped with refueling ports, which the other two spacecraft lack. This is the only area in which the good spaceship is more complex than the modular vehicle. Trans-Lunar insertion occurs well after the modular spacecraft has lost much of its complexity during staging, yet still the Lunar spaceplane is over twice as simple, and therefore twice as efficient, as the competition.

Table 10.1 Comparison of modular and spaceplane mass complexities at trans-Lunar insertion

	Apollo	Orion	Spaceplane
Fuel tanks	4	4	1
Oxidizer tanks	4	4	1
Engines	4	7	3
Attitude control systems	4	4	1
Crew cabins	2	2	1
Hatches	4	4	2
Docking ports	2	2	1
Refueling ports	0	0	2
Coupling mechanisms	4	3	0
Total	*28*	*30*	*12*

Assuming that Lunar spaceplanes are feasible, what is their real value? Once again, the answer lies in their inherent strengths as tankers and in their winged reusability. Lunar spaceplanes will transport propellants to the Moon if necessary, and from the Moon if the required resources are found. This indicates that spaceplanes will be valuable no matter which of these futures unfurls. Let us look at each aspect of this.

Lunar Infrastructure

There are three distinct uses for the Lunar spaceplane: (1) shipping propellants to the Moon, (2) shipping propellants from the Moon, and (3) transporting passengers and bulk cargo both to and from the Moon. The third use can be combined with either of the others.

The first use assumes that no Lunar ice deposits have been discovered. To keep Lunar shuttles operating between the Lunar surface and Lunar orbit, propellants will have to be transported from Earth. This will be expensive, but the alternative is to build a new Lunar spacecraft for every mission. Without a constant and reliable source of propellants to support Lunar vehicles, operations cannot be sustained. How, then, will we get propellants to the Moon? The first step is to use a series of spaceplane flights to fill a LEO depot with a propellant supply. A determination will have to be made about what types of propellant are best. The best solution is to ship ordinary water to the depot, and then to the Moon. Once there, it can be electrolyzed into its constituent rocket propellants: liquid hydrogen and LOX. The reason for this strategy has to do with the density of water compared to the density of rocket propellant. Water takes up far less volume. In addition, it is far safer to transport in its inert, nonexplosive form. Finally, transferring water from one spacecraft to another is easy and risk-free.

This raises the question, could a Lunar infrastructure be supported by simple rocket tankers instead of spaceplanes? Could such tankers be used to resupply the orbital depot, or even Lunar spacecraft directly? The answer is no, unless the rocket is also a reusable launch vehicle. Barring this, these would be one-time missions, and cost-benefit analysis exposes this idea as unworkable. The Lunar spacecraft could be reused by refueling it from a rocket tanker, while the wingless rocket tanker itself would be a worthless hulk as soon as it had delivered its propellants, and so the net gain would be zero. Rocket tankers with no wings could never reenter Earth's atmosphere and never be reused. When it comes to future space infrastructure, it is apparent that winged vehicles are just as important in space as they are on Earth.

The second use assumes that water ice *has* been discovered on the Moon, and in appreciable quantities. Well, this changes the picture completely. The Moon now becomes the key to space access itself, but again, only with an infrastructure based on winged space vessels. Lunar spaceplanes will ship Lunar ice, in the form of purified water, to LEO propellant supply depots. They will accomplish this by entering the Earth's atmosphere at an initial speed of 25,000 mph and aerobraking to 17,500 mph, relying entirely on their own aerodynamic designs and the atmosphere to achieve this delta-V. Supplies of Lunar water can be delivered to LEO "gas stations" in this manner, where they can be electrolyzed into useful propellants by solar energy. Spaceplanes arriving at these depots from the ground will refuel, and continue on to the Moon with their passengers, relief crews, and bulk cargo, returning in a few days with another tank of Lunar water. In this way, every spaceplane customer contributes to an efficient Lunar transport infrastructure.

But why ship water all the way from the Moon instead of just bringing it up from Earth? After all, the ground is *literally* a thousand times closer to LEO than the Lunar surface is, and there is a lot more water on Earth than on the barren Moon. The delta-V required to go from the Moon to LEO is negligible when compared with the ΔV needed to get from Earth's surface to an orbit 200 miles above. The most difficult aspect of spaceflight is that first 200 miles and, more important, that initial 17,500 mph. It is the difference between coasting down a long, steep hill and climbing a tall mountain. It is always easier going downhill, because gravity is doing all the work. The same is true in the case of Lunar spaceflight.

Modular Moon missions represent an extension of rocket staging, at extreme altitude. It takes three stages to get the space vehicle out of Earth orbit, and three more to reach the Moon. Modular Moon missions are therefore far more costly in terms of operational complexity than are spaceplanes. While the spaceplane simply fills up with liquid propellants in LEO and flies to the Moon, both Apollo and Orion depend on no less than six separate rocket stages, each with their own engines, tanks, interstages, propellant pressurization systems, etc. Moon missions are thereby accomplished only by throwing off each stage when it runs out of propellant and continuing on with the remaining fully fueled stages. After the three boost stages are expended and the spacecraft is on its way to the Moon, there remain three stages, or modules. These are the Service Module, the Lunar Descent stage, and the Lunar Ascent stage. In the case of Apollo, the Service Module had the task of getting the spacecraft both into and out of Lunar orbit, while the Lunar stages were responsible for getting two astronauts to the surface of the Moon and back to Lunar orbit. In the case of Orion, the large Lunar descent stage will get the spacecraft into Lunar orbit and land on the Moon. The Service Module's only job is to get the crew out of Lunar orbit for the ride home. In both Apollo and Orion, the ascent stage has one purpose only, and that is to transport astronauts and about 100 kg of rocks from the Moon to the orbiting crew module. As each stage is abandoned, the capabilities of the remaining spacecraft diminish accordingly. The system therefore has no more versatility than the capabilities of any one component. By the time the spacecraft has been reduced to conical crew-return capsule, its only capability is to protect the astronauts during the searing high-G reentry.

Spaceplanes, on the other hand, are good spaceships, because they are versatile and capable. There are no throwaway parts, and nothing is unnecessarily repeated. Because of their huge propellant tanks, they can burn their engines longer and therefore transport greater payloads, provided they can be refueled in space. And they can be, because they are part of the refueling infrastructure. The spaceplane is the space tanker for all good spaceships, including itself. It does not have to splash down under a canopy of parachutes, or rely on an armada of Navy vessels for recovery. It just lands at a spaceport and refuels, as any good plane.

Missiles and modules worked with Apollo, and the concept will work with Orion just as well, probably better. There is a small amount of reusability built into the system already. But without any resupply of propellants in space, the modular Moon mission is unsustainable. Reusable spaceships are the key, but reusability requires refueling, and refueling requires propellant transfer. Space tankers are therefore necessary for a long-term Lunar base to be feasible, practical, and safe.

The Lunar Resort

The third use of Lunar spaceplanes is transporting space tourists to their dream vacation on the Moon. Now that we have seen how the infrastructure works, it becomes a reasonable proposition to consider a Moon-based tourist economy as well. Lunar spacelines will bring along their own infrastructure as they ship propellant in the form of water to or from the Moon. In this way, they will immediately realize a double benefit in their operations. If they ship propellants to the Moon, then they can sell their liquid cargos to government astronauts, scientists, Lunar hotel shuttle operators, or whoever else needs the water. If they ship propellants from the Moon, then they can supply their own operations via Earth orbital gas stations. Either way, spacelines will turn a pretty profit, because the Lunar tourist, pretty or not, will gladly ante up his or her share.

What will a Lunar resort look like? In all likelihood, any Lunar complex will be built underground, in order to shield its inhabitants from the harmful space environment. Cosmic rays, solar flares, and coronal mass ejections are a few of the radiation hazards endemic to space. There are also extremes of temperature and the ever-present danger of meteoroids. Surface structures will therefore need to be well protected, thick-walled, or buried under the regolith. Beyond this, what might a Lunar resort have to offer its patrons?

The answer to this question depends only on the imagination and the financial resources of the owners. One intriguing possibility is the indoor soaring arena.

By attaching artificial wings to the forearm, human powered flight in the weak Lunar gravity should be a practical proposition. This will require a fairly large pressurized dome, and such arenas would not appear in the early days. But eventually, this could become very popular.

Destination Mars

Do spaceplanes have a future farther out than the Moon? We have seen how spaceplanes will serve both as tour vessels and as tankers, thereby fueling the spacelanes between Earth and Moon. Can the same logic justify a spaceplane infrastructure on the much longer, interplanetary routes? Similar to the shorter Terra–Luna runs, the interplanetary spaceways will be traveled by good spaceships that need to refuel periodically as well. The places they do this will obviously be their points of departure and arrival. We will limit the discussion to the Earth–Mars run for now, but the same logic would apply to any destination in the Solar System.

To fully understand the dynamics of interplanetary infrastructure, it is important to understand how good spaceships will arrive. Both Earth and Mars have atmospheres, which good spaceships will use for deceleration upon arrival. The atmosphere on Mars is very tenuous – only about 1% of Earth's – so it is less effective for aerobraking than Earth's own thick veil. This is not as much of a problem as it might seem, because spaceships arriving at Mars from Earth will have a relatively reduced velocity due to their uphill climb out of the Sun's gravitational well. They will be moving slower when they reach Mars than Earthbound spaceships from Mars will be moving when they reach Earth. Also, the Martian gravity field is substantially weaker than Earth's, and so Mars gravity will accelerate an arriving spaceship less than is the case with Earth. Third, spacecraft arriving at Mars encounter the atmosphere at a greater altitude, because it extends farther into space than Earth's. All of these factors tend to offset the fact that Mars's thin atmosphere seems incapable of decelerating a spaceship. The upshot of this discussion so far is that both the Martian and the Terran atmospheres are useful for slowing down arriving spaceships. This means that they must be shaped aerodynamically (Fig. 10.12).

Similar to the Lunar spaceplane, a good interplanetary spaceship will be reusable, and therefore it must be able to land anywhere and refuel. Landings on Earth require wings to generate lift before arrival at the spaceport. And landings on Mars require an aerodynamic shape for deceleration, in addition to ventral thrusters to make a VLHA landing. The alternatives dictate that the spaceship becomes some sort of partially reusable or nonreusable module, much like Orion or Apollo. This discussion concerns only good spaceships, however, which rules out modules completely.

Mars is replete with resources, both in its atmosphere and underground. Rocket propellants can be manufactured on Mars from either of these sources. When the good spaceship arrives on Mars, or in Martian orbit, the first thing it will do is fill its propellant tanks. Martian gravity is twice the Lunar, but still only a little more than one third of Earth's.

Fig. 10.12 Properly designed advanced spaceplanes will be able to aero-brake in the atmospheres of Earth, Mars, Venus, and eventually atmosphere-enshrouded moons such as Saturn's Titan (courtesy Reaction Engines Limited)

Departing from Mars, the spaceplane can use the atmosphere for lift and working mass, just as it does on Earth, even though the Martian air contains no oxygen. Two of the three components in flight and fuel management, therefore, still apply. To generate the required lift in the tenuous atmosphere of Mars requires faster speeds, but again, the gravity field is weaker, and so only a fraction of the lift required on Earth is needed on Mars. Some arrangement of air-augmented or rotating turborocket engine will also allow the Martian atmosphere to serve as working mass, just as turbofan engines regularly use bypass air for a large percentage of thrust on Earth. These considerations show that spaceplanes are beneficial in leaving the surfaces of planets such as Mars, with very thin atmospheres.

Arriving on Earth, the interplanetary spaceplane will have an atmosphere entry velocity about the same as the Lunar spaceplanes encounter, some 25,000 mph. At this point, all of the arguments in favor of the Lunar spaceplane apply also to the interplanetary spaceplane. Aerodynamic design slows the ship down, wings or lifting bodies allow the ship to navigate safely to a spaceport, and full reusability allows it to park on the ramp and refuel. It is also possible for the returning Mars spaceplane to aerobrake through Earth's upper atmosphere, reenter space on an ellipse that intersects the Moon's orbit, and land at a Lunar colony. Again, the spaceplane shows its amazing versatility and practical potential.

References

1. http://pvs.kcc.hawaii.edu/rapanui/exploration.html
2. http://science.nasa.gov/headlines/y2005/14apr_moonwater.htm
3. http://nssdc.gsfc.nasa.gov/planetary/ice/ice_moon.html
4. Harrison H. Schmitt, *Return to the Moon*. Praxis, 2006.

Chapter 11
Strategies for Success

Over the course of the next 20 years, the way we access space will undergo great changes. The space tourism market will blossom, as dozens of small companies introduce their passengers to the wonders of suborbital space travel. The majority of these companies will use simple spaceplanes that take off and land under their own power, rather than ballistic missiles or large motherships. There will be exceptions, of course. Different strategies will be tried, and different technologies will be tested. There will be successes and failures, both in terms of engineering and business strategy. A few key character traits, discussed below, will be shared by those companies who are successful in the long term. These include holding the proper vision, developing their new market, sustaining the effort, and applying lessons learned. Creativity and collaboration will play vital roles as well, because of the unique technical and financial challenges of space travel.

Establish a Vision

One of the most important factors in ensuring the success of any business is a clear and correct vision. In the space business especially, only a clear vision of the future will allow one to sustain the effort required to reach a clearly defined goal. That goal may be the eventual development of an orbital spaceplane from a suborbital one, or it may involve perching a small spaceplane atop a ballistic booster. Establishing the correct vision is at least as important as forming a clear one. If the vision is fuzzy, then goals will be difficult to set. If the vision is unrealistic, for technical, financial, or other reasons, then the goals will never be reached. The vision must be both clear and correct to ensure success (Fig. 11.1).

An example from the nascent electrical industry may serve as an object lesson. When the first electrical lines were being strung, there was debate over whether electricity should be delivered to the consumer as a direct current (DC) or as an alternating current (AC). Large sums of money were involved, and fortunes would depend on the outcome. Only one of these competing visions was correct. We now know that AC distribution systems allow the transmission of high-voltage low-current

M.A. Bentley, *Spaceplanes: From Airport to Spaceport*,
doi:10.1007/978-0-387-76510-5_11, © Springer Science+Business Media, LLC 2009

Fig. 11.1 This artist's rendering of the Skylon spaceplane cruising through the atmosphere is a good example of a clear vision of the future (courtesy Reaction Engines Limited)

power over long distances, something that is impossible for DC. Those who clung to the DC vision lost, because they had the wrong vision of the future.

In the same way, we have competing visions for the future of space access, and there are many visions at this point in history. This book presents only one of these, along with some compelling reasons for its implementation. Asking the right questions, and trying to determine the correct answers, is part of the process of making the proper choices, as the wise reader will shortly see.

Develop the Market

Space tourism will be the prime mover in pushing the future advancement of space technology and space access. This is obviously the major market for the operators of small spaceplanes. There are other markets in space, to be sure. There is always the occasional government spacecraft that needs a boost to orbit. There are scientific probes to the planets every year or so. And there are government astronauts traveling to and from space. There is the sounding rocket market. But the largest potential market, by far, is space tourism. This new market will not involve an occasional boost or a yearly probe. Instead, it will be a *daily* occurrence. With a human population approaching 7 billion, there are large numbers of people among the general population with the wealth to support this market on a continuous basis. It is the space tourist who will fund the future, one ticket at a time.

Fig. 11.2 These space pilots would no doubt feel more comfortable in the cockpit than in the passenger cabin of any spaceplane. From *left* to *right* are X-15 pilots Joe Engle (USAF), Robert A. Rushworth (USAF), John B. "Jack" McKay (NASA), William J. "Pete" Knight (USAF), Milton O. Thompson (NASA), and Bill Dana (NASA) (courtesy NASA)

Why should start-up companies focus on the space tourism market, rather than on other revenue streams such as communications satellites or space station supply? Well, the space tourism market is wide open. As long as there are customers waiting in line for the joy ride to space, the market will remain strong. Such lines have already formed, and are growing steadily. Second, the short space-experience flight has the potential to grow into a huge passenger transport industry in the future. The small space tourism company of today, which operates strictly suborbital vehicles, will be the spaceline of the future, transporting passengers between Earth and the Moon. By starting slowly and gaining large amounts of operational experience, such companies are well placed to capture the space passenger markets of tomorrow.

Sustain the Effort

To reach the goal of eventually providing regular passenger access to orbital or Lunar space, the small spaceplane development company of today must embark on a slow, steady program of incremental improvement in design and technology. Already, some companies are advertising flights to 130 km, while others are expecting a suborbital apogee of nearly three times that height, with resulting

Fig. 11.3 Teacher-in-space Christa McAuliffe experiences weightlessness in NASA's KC-135 zero-G training aircraft prior to her ill-fated mission aboard the *Challenger*. This clear vision will help to develop the imminent space tourism market (courtesy NASA)

longer periods of weightlessness. As suborbital flights gradually lengthen their zero-G endurances and raise their apogees, one will eventually reach orbital velocity itself. But it will require a sustained effort. It may take decades and involve the invention of new, lightweight, combined cycle engines such as the turborockets we discussed in Chap. 8.

Patience and perseverance are the twin traits of the tortoise who won the proverbial foot race, as well as of the team who wins the private space race. Unlike the tale of the tortoise and the hare, there will be many more winners, and many more losers. An ongoing incremental program of engine and airframe improvement will eventually lead to the first SSTO spaceplane. Along with these twin character traits will be the necessity of maintaining a strong faith in a concept, in a company, in a team, and in a vehicle.

Apply Lessons Learned

A thorough study of all past failures will separate the successful company from the one that falls by the wayside. Every lesson, whether it be of a technical, financial, or human nature, must be studied, learned, and applied. Failure to study the lessons of the past is a failure in itself, and those who fail in this basic nugget of common sense are only dooming themselves to their own failure. One of the most important lessons is to study past space efforts, and see whether they worked, why they worked, how well they worked, or why they did not work. If there is a pattern of concepts that were considered but repeatedly canceled, it would behoove the smart company to figure out why. This lesson can help the wise establish the proper vision for the future. The company with the wrong vision, and the wrong goals, will fail, just as assuredly as the DC power distribution network was doomed from the beginning.

Ask the Right Questions

Part of this process is knowing which questions to ask, and this requires vision also. Asking the right questions is a good thing, but answering them correctly is even better. Here is a short list of a few basic questions, together with my answers. These questions already assume that you have already chosen spaceplanes over ballistic capsules or wingless launch vehicles.

- *Should I perch my spaceplane on top of my rocket?*

No – too much dead weight. Your booster will be too expensive to operate, unless you are already rich.

- *Should I mount my spaceplane piggyback?*

No – too costly and operationally complex. Again, you will go broke before you get off the ground (Fig. 11.4).

- *Should I use a mothership?*

Fig. 11.4 It is time for piggyback spaceplanes to grow up (courtesy NASA)

Fig. 11.5 Only baby spaceplanes are carried by their mothers. This is the X-1-3 being carefully mated to an elevated EB-50A Superfortress. A typical government operation (courtesy NASA)

Fig. 11.6 The Douglas Skyrocket was capable of taking off under its own power in certain configurations. It therefore represents the proper approach to mature spaceplane development (courtesy NASA)

No – again, this is too complex and costly in the long run, because you have to operate two vehicles (Fig. 11.5).

- *Should I let my spaceplane fly on its own?*

Yes. This way it can mature into a proper SSTO spaceplane in time. Like the kid that never had training wheels, this little spaceplane will develop superior capabilities by itself (Fig. 11.6).

- *What kinds of propellants should I use?*

Liquid hydrogen and liquid oxygen – water electrolyzed. This way you can ship water through space, and someday refuel there. This answer requires clear vision.

- *What is wrong with easy-to-use-fuels such as kerosene or methane?*

With these fuels you cannot refuel in LEO, and that cancels the future. Also, they pollute.

- *Should I lift off vertically or take off horizontally?*

Horizontally – the main reason is passenger safety and comfort, but it also enables conventional operations.

- *Should I operate my crewless spaceplane by remote control?*

Remember DC-X and X-33? Besides, who wants to ride in a pilotless spaceplane?

- *What kind of engine should I use?*

The most advanced air-breathing turborocket you can find, assuming good reliability.

- *I do not have a turborocket handy. Should I use jet engines in addition to my rocket engines?*

Yes. Auxiliary jets will increase safety and enable you to light off from high altitude. A single crash without jets installed, and you are finished. Jets will increase ticket sales, too.

- *What is the single most important factor in my spaceplane design?*

Safety, which is the reason you chose wings to begin with.

Be Creative

Creative ideas may mean the difference between success and failure. Space business and space technology together make up a brand new game. It is certainly not business as usual, and new ways of thinking will be necessary. This new mentality will involve Web-based relationships, bold strategies, and new technologies. Bold, innovative leadership and perseverance in the face of ridicule will also be required. It should be loudly repeated, over and over, that it is possible to access space from a runway instead of a launchpad. Gradually, the flying public will begin to absorb this new knowledge.

Creative solutions are often better than more traditional ones, although tradition will play an important role in the spaceplanes to come. One such tradition should be the dogged insistence on horizontal takeoff and landing. The missiles and modules idea should be cheerfully chucked in favor of the fully reusable, completely reliable, common-sensical spaceplane.

Collaborate!

The Internet has opened up a whole new world in terms of collaboration. Long-distance cyberspace friendships can be of real value when it comes to entering near-Earth space. Communicating through the World-Wide Web was never easier, and such communication has never been cheaper. This factor alone means that collaborations, especially international ones, will be possible like never before. Snippets of information can be instantaneously transferred between continents, enabling the space entrepreneur to share knowledge and ideas at light speed. Those who jealously guard their ideas will only be hurting themselves. Lessons can be taken from both the Wright brothers and Goddard, in this respect (Fig. 11.7).

The high costs of aerospace technology development in the early years of the twenty-first century virtually dictate collaborations for financial reasons alone,

Fig. 11.7 Healthy collaboration is important in any venture. Of course it also helps if you know someone rich. Shown are Walt Disney (*left*) and Wernher von Braun (courtesy NASA)

unless the company is graced by some kind of financial independence or government contract. The alternative may be stagnation for a venture with otherwise workable ideas. The bottom line is funding. Without it, nothing happens, and no one flies. Even Robert Goddard relied on backing from the Smithsonian Institution to conduct his rocket research.

Deliver the Package

Spaceplanes today are in an infantile stage of development. As we have seen, they all either ride piggyback or rely on the nestling wing of a large mothership. But they have a lot of development potential. They could soon replace sounding rockets, reducing costs in this market by immediately introducing full reusability. They will shortly take passenger-tourists on space experience flights above the official 100 km space height. This event, the first suborbital spaceplane flight with a paying passenger, will open the floodgates. There will be a surge of development as investors begin to realize the potential. As they continue to grow and improve, spaceplanes will eventually become capable of orbital spaceflight, without the assistance of an auxiliary booster, drop-off stages, or a carrier aircraft. At this point, they will be pressed into service as space tankers, transporting pure water to orbiting propellant depots, to the Moon, and someday from the Moon. All the while, they will continue to provide regular service to a wider and wider array of space travelers. Spaceplanes will deliver the package, and do it far more cheaply, far more efficiently, and far more safely than will primitive ballistic rockets.

Fig. 11.8 Infant spaceplane prototype, the Spiral spaceplane in Soviet colors. CCCP is Russian for SSSR, which stands for *Soyuz Sovietskikh Sotsialisticheskikh Respublik* (Union of Soviet Socialist Republics) (courtesy http://www.buran-energia.com)

To deliver the space packages of the future, whether these packages take the form of bulk cargo, liquid propellants, or passengers, good spaceships are required. Missiles and modules will never become good spaceships, regardless of how far they evolve. Spaceplanes, by contrast, have a very real potential to grow into the best spaceships of all.

Chapter 12
Spaceplanes at the Spaceport

Today we are living on the cusp of the one of the greatest changes in human history. We are climbing out of our planetary cradle, using chemically fueled rockets, as Konstantin Tsiolkovskiy clearly foresaw so many years ago. Although large bureaucratic boosters have accomplished amazing feats, the privately owned spaceplanes now being developed in several countries are, even now, writing the prologue to all future space-based civilization. It is they, and their descendants, who will take humankind by the millions into the vacuous voids of interplanetary and, someday, interstellar space.

Today's Spaceplanes

Following is a short run-down on current projects, as they stand in the first decade of the twenty-first century.[1] This list is by no means exhaustive and is in a constant state of flux.

Benson Dream Chaser

Benson Space Company based in Poway, California, has a new design, the Benson Dream Chaser. It uses a sleek and simple rocket, launches vertically on hybrid motors to take a small winged rocketship to space height, and lands on a runway. Dream Chaser's aerodynamic design requires less propulsion and promises G-forces less severe than the competition. Passengers will pay $200,000 to $300,000 per ride. The idea is to lift off vertically in a sort of miniature version of a NASA rocket launch, and then land like a minishuttle. The little rocketship runs on hybrid motors similar to the ones that powered SS1 but does not require a mothership. Propellants are liquid nitrous oxide (N_2O), the oxidizer, and solid HTPB (hydroxyl-terminated polybutadiene), the synthetic rubber fuel. Dream Chaser is right on target for its first commercial flight in 2009 and is currently taking reservations.

M.A. Bentley, *Spaceplanes: From Airport to Spaceport*,
doi:10.1007/978-0-387-76510-5_12, © Springer Science+Business Media, LLC 2009

Bristol Spaceplanes Ascender

The horizontal take-off and landing Ascender (Fig. 12.1) is the brainchild of aerospace engineer David Ashford of Bristol Spaceplanes, Bristol, England. He is also the author of one of the first books about spaceplanes, *Spaceflight Revolution*, published in 2002 by Imperial College Press. The initial design calls for a single stage suborbital spaceplane that will take off under its own jet engines from a conventional runway, and rocket into space in a steep ascent trajectory. It will then glide back to Earth and land at the point of departure. Future designs call for the mating of Ascender with a large HTHL booster aircraft which will take the orbital vehicle up to supersonic speeds in the upper stratosphere. The winged orbiter will then continue into space while the winged booster returns to its runway for reuse. There is more information on this design in Chap. 7.

EADS Astrium Spaceplane

The European Aeronautic Defence and Space Company's Astrium Space Transportation division announced, in June 2006, that they would build on the concept of the *Rocketplane XP,* as the most viable commercial spacecraft concept. This huge aerospace company, second only to Boeing, studied this and other vertical launch concepts before making its determination. Its endorsement of the spaceplane idea is therefore a significant development.

The business jet-sized Astrium spaceplane will take one pilot and four passengers on 90-min. cruises into near-Earth space above 100 km. Similar to *Rocketplane XP*, the EADS spaceplane will take off and land on runways using conventional jet

Fig. 12.1 The Bristol Spaceplanes Ascender pictured above the Earth's atmosphere. Another Ascender can be seen in the background (courtesy Bristol Spaceplanes)

engines. These will take the spaceplane up to an altitude of some 39,000 ft, at which point the methane-fueled rocket engine will ignite and burn for 80 s. This boosts the Astrium spaceplane to 200,000 ft on autopilot, followed by 3 min of Zero-G ballistic flight peaking at or above 330,000 ft. Small rocket thrusters maintain control during the airless, weightless portion of the trajectory. Passengers are expected to be able to float out of their seats for several minutes before buckling back in, prior to atmospheric reentry.

A unique feature will be balancing, self-adjusting seats to ease the 4.5-*G* reentry loads on space tourists. For $175,000 to $220,000 each, passengers will go through 1 week of preflight training, as opposed to Soyuz passengers who endure 6 months of strenuous training after paying at least 100 times that amount. The EADS spaceplane is expected to take 4 years to develop beginning in 2008, with first commercial spaceflights in 2012, given adequate private investment estimated at $1.3 billion. Private consultants expect a market of 15,000 space tourist passengers per year by 2020.[2]

Ramstar Orbital Spaceplane

Aerospace Research Systems, Inc. (ARSI), of Withamsville, Ohio, is developing its own HTHL spaceplane under the leadership of Dr. Pamela Menges, Ph.D. This unique design has already completed in-flight testing of some 80% of its advanced systems. The Ramstar (Fig. 12.2 and 12.3) will utilize an "artificial neural membrane" (ANM). The ANM technology is a new class of functional structures based on a new materials-manufacturing technology, whereby the structures themselves possess resident information processing, communications, sensing, and control systems. This technology is central to the design and operation of the Ramstar, with an ANM-based "smart skin," adaptive jet engine intakes, optronic (light-based) computing systems, and advanced life support and payload systems. ARSI smart skin technology, with the thickness of about three sheets of paper, is also being used to achieve active drag reduction using electromagnetic fields. Furthermore, microwave spikes may be used in the future to reduce heating along leading edges.

The multirole two-stage-to-orbit Ramstar will take off from a runway powered by onboard jet engines, and activate externally mounted boosters at altitude to accelerate spaceward. The vehicle may be flown either autonomously or with a crew. The highly flexible design uses a modular cockpit, modular payload bays, and minilabs that slide in on rails. This allows payloads 18 m in length and 18,800 kg in mass on autonomous missions or 12 m and 12,000 kg with crew and modular flight deck installed. The craft will sport a double-layered titanium hull with passive high-temperature coatings in addition to the active drag reduction systems. The flying testbed now nearing completion will be a scaled down version of the full-size model. The reusable booster rockets initially will use either a conventional solid propellant or a solid–liquid slurry system now being developed. These boosters are specifically designed to shut down in flight if necessary, enabling increased flexibility, greater convenience, and enhanced safety. The plan is to manufacture the Ramstar in two versions, a smaller vehicle with a turn-around time of only about

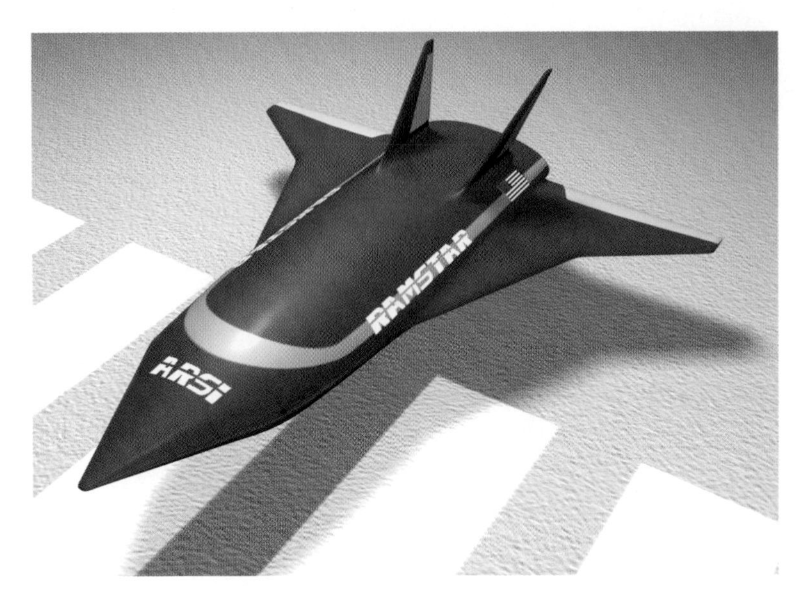

Fig. 12.2 Artist's impression of Ramstar Hypersonic Unmanned Aerial Vehicle shown about to depart from a conventional runway (courtesy Aerospace Research Systems, Inc.)

Fig. 12.3 The Ramstar Orbital Spaceplane may utilize a boron slurry in a wide-range ramjet, or use a combined-cycle turbojet/scramjet or hypersonic wave engine (courtesy Aerospace Research Systems, Inc.)

4 weeks, and a larger version with a turn-around time of 6–8 weeks. Both can be flown either with or without human crews aboard. Up to a dozen of each version could eventually be built, with production scheduled to begin in 2009.

Rocketplane XP

Rocketplane Global, based in Burns Flat, Oklahoma, is currently taking reservations for suborbital spaceflights on the *Rocketplane XP*. The company has already spent 10 years working on the design. The *XP* spaceplane will carry one pilot and five passengers on suborbital rides, beginning in 2010, using a conventional-looking airplane with dual afterburning J-85 turbojets and a single rocket engine. One passenger will pay $250,000 to ride in the copilot's seat, while the other four will pay $200,000 each to be regular passengers.

The *XP* will not rely on a mothership but will leave the runway like any airplane, powered by its own jet engines. The afterburning turbojets will allow the *XP* to reach an altitude above 40,000 ft before the rocket engine has to be ignited, thereby lengthening the amount of zero-G time and enhancing the suborbital spaceflight experience. It will incorporate a T-tail at the rear of the 44-ft fuselage and weigh only 20,000 lb at take-off. Its single Rocketdyne RS-88 engine will generate 36,000 lb of thrust, providing a good thrust-to-weight ratio for the nearly vertical climb into suborbit. The trajectory will take the *XP* beyond 330,000 ft, the official space demarcation line. *Rocketplane XP* will burn liquid oxygen and kerosene in the rocket engine, and kerosene in the turbojets. Maximum speed will be just over 3,500 ft/s, or about 2,400 mph at rocket engine cut-off. Upon return to Earth, the jet engine will power back up at about 30,000 ft and bring the *XP* spaceplane back to the same spaceport that it took off from. The eventual goal is to one day reach the developmental stage of the advanced orbital spaceplane.[3]

Skylon

Certainly one of the sleekest designs of all is the Skylon advanced spaceplane concept of Reaction Engines Limited, based in Oxfordshire, England. It is also one of the largest. The greater part of the volume of Skylon is taken up by propellant tankage, precisely what is required to achieve single-stage to orbit capability in a reusable spaceplane. Skylon is capable of flying without a crew, which increases available room for propellants. Its advanced combined cycle turborockets enable the craft to deliver 12 metric tons of cargo directly to low Earth orbit without wasting spent stages. It makes full use of the atmosphere, using both its wings and its advanced engines to enter orbit (Fig. 12.4).

Skylon's slender fuselage contains huge propellant tanks for the liquid hydrogen and liquid oxygen, as well as a sizable payload bay designed to accept standard launch packages. Delta wings support twin dual-mode Sabre turborocket engines in axisymmetric nacelles mounted on the wingtips. Foreward of the main wings are canards for pitch control in the atmosphere. An all-movable vertical stabilizer and conventional ailerons provide yaw and roll control. In space, a reaction control system draws from the main propellant supply. The vehicle operates from reinforced

Fig. 12.4 Skylon spaceplane in low Earth orbit. It requires no external tanks, drop-off boosters, stages, or carrier aircraft. It is the epitome of the advanced spaceplane (courtesy Reaction Engines Limited)

Table 12.1 Skylon specifications

Length	82 m	270 ft
Diameter	6.25 m	20½ ft
Wingspan	25 m	82 ft
Empty weight	41,000 kg	90,000 lb
LH_2 fuel	66,000 kg	145,000 lb
LO_2 oxidizer	150,000 kg	330,000 lb
Maximum payload	12,000 kg	26,000 lb
Max. takeoff weight	275,000 kg	600,000 lb
Landing weight	55,000 kg	120,000 lb

runways without auxiliary aircraft or boosters. It is a completely self-contained vehicle, landing and taking off like any airplane.

The Sabre engine, also designed by Reaction Engines Limited, can operate either in an air-breathing or in a rocket mode. From takeoff to about Mach 5, it draws in atmospheric air, to be used as oxidizer and propellant, by means of a two-shock axisymmetric intake. The air is precooled via a closed helium loop just before entering the combustion chamber. The helium loop is itself cooled by the cryogenic liquid hydrogen fuel. Therefore the helium is never used up while performing useful work. In rocket mode, the engines use the onboard liquid oxygen and liquid hydrogen (see also Fig. 12.5).

There is additional information on Skylon and Sabre at http://www.reactionengines.co.uk.

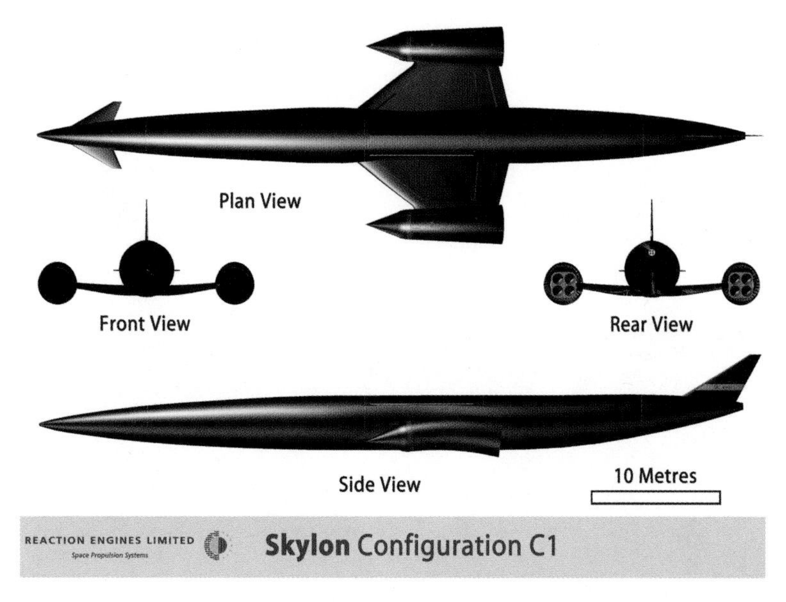

Fig. 12.5 General three-view arrangement of the Skylon advanced spaceplane concept (courtesy Reaction Engines Limited)

Spacefleet SF-01

The private British SF-01 (Fig. 12.6) is in many ways the most intriguing of the currently contemplated spaceplane concepts. It is the largest design among the suborbital vehicles, carries the most passengers, costs the least to develop, charges little, is cheap to operate, and does not pollute the planet. It is still early in the maturation process and will likely evolve over time, but here are the basic facts as of October 2007.

The Spacefleet Project, based in southern England, plans to build a fleet of three environmentally friendly spaceplanes powered by liquid hydrogen and liquid oxygen. Not only are these the best rocket fuels around, but they can be produced by electrolysis of water using a solar photovoltaic array and/or wind energy. In this way, producing the propellants is a perfectly clean process, as is burning them in the engines, since the only combustion product is water vapor. The Spacefleet Project claims that its spacecraft is the only one under development known to inject zero carbon emissions into the atmosphere. Furthermore, production of these clean, energetic propellants is free and sustainable once the equipment is up and running.

The SF-01 has four LOX/LH_2 four-chamber rocket engines with a mixture ratio of 7 to 1, a thrust per engine of 150 kN, a specific impulse of 350 s, and a burn time of 125 s. The titanium hull measures 14 x 14 x 3 m, with the height increasing to 6 m when the landing gear is lowered. The final design may include turbojet engines, but this is still under review. SF-01 carries 56 m^3 of high-energy propellants, enabling it

Fig. 12.6 The SF-01 shown as it might appear at 100 km altitude. The large lifting body design has room for large propellant tanks and a spacious passenger cabin (courtesy Spacefleet Ltd.)

to take no less than eight passengers and two pilots to an apogee of 340 km after launch from an inclined ramp at the spaceport. Passengers will experience a maximum of 2.4 *G* during the 2-min burn of the rockets, which cut off at 140 km. They will enjoy a full 6½ min of weightlessness as the spaceplane soars to the same altitude as the International Space Station. This is nearly three times the height reached by other space tourist vehicles, a factor that should increase passenger excitement and satisfaction significantly. Similar to other spaceplanes, it will glide back and land on the spaceport runway after each spaceflight. The SF-01 is also able to make intercontinental "point-to-point" flights, deploy small satellites, or carry scientific payloads in its large payload bay, with appropriate modifications (see also Fig. 12.7).

The development project is expected to take 3 years and cost no more than €260 M, finishing with delivery of three vehicles. The electrolysis plant should cost another €150 M, but thereafter should be cheap to operate. The company is now actively seeking funding.

Spacefleet is well-placed to gain the experience necessary for future orbital refueling operations, one of the keys to future spaceflight. The project scientist and managing director is Dr. Raymond D. Wright, an electrochemist with a quarter-century of experience in the British electrical industry and founder of Spacefleet, Ltd. He is therefore well qualified to head up the project. Ticket prices are projected to be a competitive $120,000, little more than half what much of the competition is charging.[4]

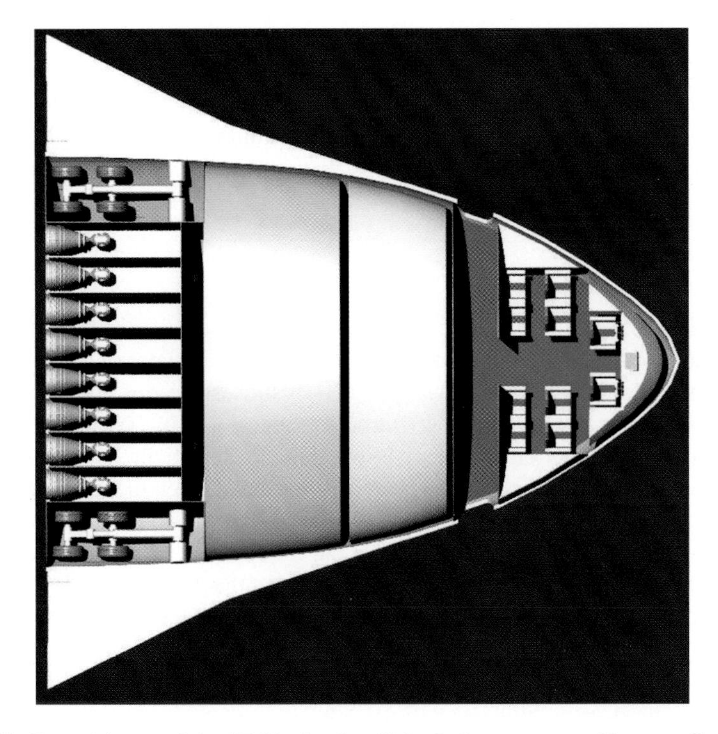

Fig. 12.7 General layout of the SF-01, showing flight deck, passenger cabin, propellant tanks, landing gear, and engine bank (courtesy Spacefleet Ltd.)

SpaceShipOne

The first privately financed, privately designed, and privately built spaceship was the spaceplane SpaceshipOne. Its first powered flight took place on December 17, 2003, exactly 100 years after the Wright brothers made the first powered heavier-than-air flight at Kitty Hawk. Between June and October of 2004, SpaceShipOne made three suborbital spaceflights, each exceeding 100 km altitude. The last two flights took place within 2 weeks of each other, winning the X-Prize for Scaled Composites, builder of the baby spaceplane.

SpaceShipOne made only three flights into space before being sent to the National Air and Space Museum in Washington, D.C. to take its rightful place among other famous trailblazing craft in aerospace history. It now hangs with the Bell X-1, the first aircraft to break the sound barrier, and *The Spirit of St. Louis*, the first airplane to cross the Atlantic Ocean.

SpaceShipOne proved that a private company could design, engineer, build, and fly a spaceplane into suborbit, and repeat the feat within 2 weeks. Similar to its predecessor the X-15, SpaceShipOne was carried under the wing of a mothership and released at high altitude. Unlike the X-15, this new-generation baby spaceplane used hybrid rocket engines utilizing a solid fuel and liquid oxidizer. This enabled the spaceplane to control its engine, as others of liquid propellant design.

SpaceShipTwo

SpaceShipTwo, the next-generation successor to the privately funded SpaceShipOne, promises to carry two pilots and six passenger-tourists to altitudes of some 75 miles, finally surpassing the altitudes flown by the X-15 almost 50 years ago. Three times larger than SS1, the SS2 spaceplane will be air-launched at 60,000 ft by the White Knight 2 mothership, currently under construction. White Knight 2, also three times larger than the first White Knight, has been christened *Eve*. Each flight will take about 2½ h and cost passengers around $200,000. When the powered portion of the flight ends, passengers will be able to float about the cabin for a few minutes. They may be tethered to their form-fitting seats, however, so that the pilot can "reel" them back in before reentering the atmosphere. Testing of SS2 will take place at the Mojave Spaceport in southern California, with commercial flights departing and arriving at Spaceport America, New Mexico, starting in 2009 or 2010.

The next phase, SpaceShipThree, is expected to be capable of orbital operations, and may dock at either the ISS or a private space station. The launch architecture for SS3 has not yet been unveiled.

XCOR Xerus

XCOR Aerospace, based at Mojave Spaceport, California, is developing the versatile Xerus suborbital spaceplane, to be powered by in-house rocket engines featuring XCOR piston pumps. These new pumps promise greater reliability than conventional turbopumps. Xerus will take off from a spaceport runway under its own power, climb to 65 km, and cut its engines. It will then coast to 100 km, where it will conduct one of three missions: (1) microgravity research, (2) microsatellite launch, or (3) space tourism. Maximum velocity will be Mach 4. Restart of the rocket engines, with the help of quick-acting piston pumps, will make landings safer and go-arounds possible. The Xerus will utilize liquid propellants only, either methane or kerosene as the fuel and liquid oxygen as the oxidizer. Methane (CH_4) offers great promise as a rocket fuel, because it can be manufactured in places such as Mars, or pumped directly from the surface of Titan at some point in the future. It is also easily liquefied for use in spaceplanes. XCOR plans to charge tourists $98,000 for brief space experience flights of at least 30 min, including 3 min of zero gravity, starting in 2009 or 2010.[5]

Space Hotels

Once spaceplanes have reached the stage at which they are capable of orbital operations, the space hotel will be their natural destination. Wealthy space entrepreneur Robert Bigelow has both the means and the desire to see this kind of future unfurl – or in his case, inflate. His space hotel designs are essentially portable, inhabitable balloons, delivered to low Earth orbit in compact packages by conventional rocket. Once inflated, they promise to provide the luxurious accommodations expected by future space tourists. For those with the monetary means, the orbital sojourn promises to deliver the experience of a lifetime at "bargain" rates. The weekly charge in the Bigelow Sundancer, for example, is expected to be a reasonable $8 million. The Sundancer will provide 6,400 ft^3 of floating space and three windows, to accommodate a crew of three and three space guests. The proto-hotels Genesis I and Genesis II have already been launched into low Earth orbit, the latter with a cadre of minuscule organisms, in order to pave the way for future human visitors.

Airports Today, Spaceports Tomorrow

Every major airport on Earth today will at some point in the future become a spaceport. These spaceports will be upgraded versions of today's international airports, because all arriving and departing spaceships will do so as spaceplanes, utilizing the atmosphere for propulsion and lift the same as any airplane. For a long time to come, most spaceports will be on Earth, for the simple reason that most people live on that planet. But spaceports will be built on other bodies as well, starting with the Moon.

The Lunar spaceport will not resemble any spaceport on Earth, because the Moon lacks an atmosphere. As a result, runways will be conspicuously absent. All spacecraft will arrive and depart vertically, using landing and ascent rockets. One feature the Lunar spaceport may have that the typical Earth facility will not is an elevator. This will lower the recently arrived spaceplane into a pressurized underground environment, where passengers and crew may disembark without the need to first don bulky spacesuits.

Tomorrow's Spaceplanes

The main thrust of this book has been a description of the spaceships of tomorrow, which ought to be spaceplanes. As we have seen, the chief reason for the development of spaceplanes lies in their inherent reusability, regardless of whether they are meant for suborbital, orbital, or Lunar flights. Reusability is the key to the future, and wings are the means by which that key is forged. Without wings, we are stuck with egregiously expensive ballistic boosters, blunt reentry capsules, and infrequent space shots. Fixated on rockets, we remain mired in a morass of missiles and modules.

Spaceplanes will free us from the bonds of conventional thinking in terms of space access. They will finally allow us, the willing and freedom-loving denizens of planet Earth, the chance to enter the vacuous void and float unfettered in that fantastic fishbowl called the universe. This will come about as an inevitable result of drastically lowered costs when fully reusable spaceplanes begin plying the interorbital spaceways. As has been true throughout human history, wealth and its financial tools have always driven exploration and development. Therefore, cost-effective spaceplanes, transporting passengers, cargo, and propellants, and thereby filling simultaneous, vital, and multiple niches in future space infrastructure, will enable a future of expanding human activity centered on Earth. Winged space vessels promise vastly superior performance to any other type of spacecraft yet unveiled. The preference for the spaceplane touches many areas of human development, from the economic, to the technological, to the personal. Whether you own and operate a fleet of spaceplanes, or you simply wish to ride in one as a passenger, the benefits of the spaceplane are obvious. And yet, after spaceplanes have waxed and waned from the human scene, other, far more fantastic ships will rise from Earth.

Beyond the Spaceplane

Someday the spaceplane will be as fondly remembered as the Conestoga wagon of the American pioneer, or the nineteenth-century steam locomotive. Space history books of the future may actually devote an entire chapter to the ancient spaceplane, the first fully reusable, practical spaceship. What type of space vessel could possibly succeed the spaceplane?

Certainly, anything more advanced than the spaceplane will have all the features that ensured its own success: reusability, ease of operation, affordability, reliability, and, of course, safety. What, aside from magic, could upstage these characteristics? To be sure, the fantastic craft beyond the spaceplane will be rooted in technology as firmly as any other machine. What might these features be? Here we enter the realm of informed speculation, liberally laced with a dose of creative imagination, yet still rooted firmly in factual physics.

The space vessels of the extreme future will not require wings at all, nor will they even require propellants. They will, nevertheless, harness far greater energies than any rocket or spaceplane that came before it. They will be capable of not only interplanetary flight on a routine basis but also interstellar travel. Their energy source and means of propulsion will be one and the same, based on an understanding of gravitational field physics that our best twenty-first-century scientists have yet to master. With this knowledge, engineers of the future will design spacecraft that will create local distortions in gravity fields, enabling extreme accelerations without inertial effects. Advances in physics, likely supplanting or updating Einstein's theory of General Relativity, will inevitably enable the construction of these seemingly magical devices. By manipulating gravity, both space and even time itself will be controlled, allowing our species the freedom to finally roam the universe. But that is the subject of another book!

References

1. http://www.space-tourism.ws/index.html
2. http://edition.cnn.com/2007/TECH/space/09/27/euro.space.tourism/index.html
3. http://www.rocketplaneglobal.com/index.html
4. http://www.spacefleet.org.uk/project/sf-01.htm
5. http://www.xcor.com

Appendix A
Glossary

Accelerometer
Device that measures the acceleration of a rocket or space vehicle, used in inertial guidance systems. In an integrating accelerometer, the acceleration is integrated over time to give accurate velocity and position readouts.

ACS
Attitude control system, the system of small thrusters that allow a spacecraft to control its attitude in the vacuum of space.

Aerobraking
The technique of using a planet's atmosphere to decelerate a spacecraft by aerodynamic friction, rather than having to use onboard propellants.

Aerospike
An "inside-out" rocket engine, incorporating a typically truncated linear or annular "ramp" that is exposed to the open atmosphere. The exhaust gases are expelled into the region between the ramp and the atmosphere, resulting in continuous altitude compensation (see separate entry), optimum expansion, and increased efficiency at any altitude or pressure.

Altitude compensation
The automatic and continuous optimum expansion of the exhaust gases exiting a rocket nozzle in reaction to changes in ambient pressure conditions. Conventional nozzles allow optimum expansion at only one ambient pressure, resulting in overexpansion (see separate entry) at lower altitudes and underexpansion (see separate entry) at higher altitudes.

Ambient pressure
The local prevailing atmospheric pressure.

Antipode
The point in a ballistic trajectory halfway around the world from the launch site.

Apogee
The point of highest altitude in an elliptical orbit or suborbital trajectory.

APT
Aerial propellant transfer.

Area ratio
The ratio of a rocket nozzle's exit plane area to its nozzle throat area.

Aspect ratio
Wingspan (b) divided by average chord (c), also equal to the square of the span divided by the wing planiform area (S). Stubby wings have a low aspect ratio.

$$A = \frac{b}{c} = \frac{b^2}{S}$$

Attitude
The orientation in space of a flight vehicle relative to some reference frame.

Ballistic coefficient
A determination of how much heating takes place during atmospheric reentry.

Booster
A high-thrust rocket stage used to help lift heavy payloads into orbit or boost spacecraft further into space, usually dropping off after use. Many modern launchers use solid rocket boosters in the initial stage of flight.

BOR
Bespilotniy Orbitalniy Raketoplan, the Russian acronym for unpiloted orbital rocketplane.

Boundary layer
The thin layer of fluid (gas or liquid) between a vehicle surface and the free stream. The molecules nearest the surface adhere to the surface and those adjacent to the free stream move with it. A boundary layer may be laminar or turbulent, and may be assumed to be viscous (gluey) or inviscid (frictionless).

Carbon vane
A jet vane made of carbon, placed in the rocket engine exhaust stream for steering control.

Chord
The straight-line distance between a wing's leading and trailing edges.

CM
The Apollo Command Module; also, the Orion Crew Module.

Combustion chamber
The enclosed area in a thrust chamber (see separate entry) in which rocket propellants are ignited and burned before accelerating through a convergent–divergent throat and expanding through a nozzle.

Congreve, Sir William (1772–1828)
A British lieutenant colonel who developed working solid-fuel one- and two-stage rockets, time fuses, and parachute systems for the Royal Army. Congreve's rockets are immortalized in *The Star Spangled Banner*, the national anthem of the United States.

CSM
Command and Service Module (Apollo); also, Crew and Service Module (Orion).

Cryogenic propellant
An extremely low temperature liquid rocket fuel or oxidizer. Examples are liquid hydrogen (–435 to –423°F) and liquid oxygen (–362 to –297°F).

De Laval nozzle
Conventional convergent–divergent rocket nozzle, named for the Swedish engineer Carl de Laval (1845–1913). The shape accelerates exhaust gases to supersonic speeds at the nozzle exit.

Delta-*V*
The theoretical change in velocity, or ΔV, expected from a rocket burn in the absence of performance losses due to gravity, atmospheric drag, back pressure, etc. Specific impulse (see separate entry) and mass ratio (see separate entry) are the main factors that determine ΔV.

$$\Delta V = c \ln\left(\frac{M}{m}\right)$$

Drag loss
The reduction in velocity increment (see separate entry) affecting launch vehicles operating in an atmosphere, due to aerodynamic drag.

Dynamic pressure
The pressure q felt by a flight vehicle due to the velocity v of onrushing air molecules with density ρ

$$q = 2\rho v^2$$

Eccentricity
The ratio e of the distance between the foci ($2c$), to the length of the major axis ($2a$) in an ellipse, a hyperbola, or a parabola (see separate entries).

$$e = \frac{c}{a}$$

EDS
Earth Departure Stage, part of the Ares V launch vehicle.

Ellipse
An oval-shaped curve, such that the distance from one focus to any point on the curve and back to the other focus remains constant. All ellipses have eccentricities

(see separate entry) between 0 and 1, and all closed orbits are ellipses, with the parent body occupying one of the foci.

$$\frac{x^2}{a^2} + \frac{y^2}{b^2} = 1$$

In this standard equation, the center of the ellipse lies at the origin of the Cartesian coordinate plane. The ends of the major and minor axes are located a and b units from the y and x axes, respectively, and each focus lies on the x-axis at a distance $c = \sqrt{a^2 - b^2}$ from the origin.

Envelope
The particular range of altitudes and airspeeds characteristic of a given aircraft. Spaceplanes will clearly require much larger flight envelopes than any other aircraft.

Escape velocity
The speed required to permanently escape from a gravitating body in space. It is always equal to the square root of two times the circular orbital velocity. In the following expression, g refers to the local acceleration of gravity at the spacecraft's location, and r is the radial distance from planet's center to the spacecraft.

$$V_{\mathrm{esc}} = \sqrt{2gr}$$

Exhaust velocity
The speed (c) that exhaust products depart the nozzle exit plane, a critical factor in determining specific impulse (I_{sp}) and rocket performance. It can be found by taking the ratio of thrust (F) to propellant mass flow rate ($\dot{m}$).

$$c = I_{\mathrm{sp}} g_{\mathrm{e}} = \frac{F}{\dot{m}}$$

FFM
Flight and fuel management.

Film cooling
A method of cooling a rocket thrust chamber from the inside, using a thin film of propellants injected against or very near the chamber walls. As this film vaporizes, it forms a cool layer next to the chamber walls, providing effective cooling.

Free fall
The state of weightlessness characterized by a body falling freely in a gravitational field. Since gravity fields pervade all space, orbiting objects are in free fall around the central body.

Free stream
The stream of fluid moving relative to a vehicle operating in air or water, and typically isolated from the vehicle by a boundary layer (see separate entry).

Fuel
Solid, liquid, or gaseous propellant (see separate entry) that produces useful heat or energy when burned with an oxidizer (see separate entry). Examples are ethyl alcohol, liquid hydrogen, kerosene, methane, HTPB, and powdered aluminum.

Gantry
A launch vehicle service tower, used to provide access to launch team workers, electrical connections, and propellant lines.

Gas generator
A device used to generate and deliver hot pressurized gases or steam to other parts of a rocket propulsion system. These gases are typically used to pressurize propellant tanks or spin the turbine of a turbopump.

GEO
Geosynchronous earth orbit, an orbit at an altitude of 22,300 miles in which a satellite seems to hover over one location on the ground, typically used by communications and meteorological satellites.

***G*-force**
An inertial force that mimics weight and is experienced by astronauts during periods of high acceleration or deceleration.

Gimbal mount
A pair of connected swiveling frames hinged on mutually orthogonal axes, allowing a thrust chamber to point in any direction, thereby enabling thrust vector control. Gimbal mounts are also used to support gyroscopes and gyroscopic flight instruments, which maintain fixed attitudes in space while the structure of the vehicle itself tilts or rotates.

GLOW
Gross lift-off weight; also, gross light-off weight.

Goddard, Robert H. (1882–1945)
An American physics professor and pioneer rocket scientist. On March 16, 1926, Goddard flew the first liquid-fueled rocket from a farm near Auburn, MA, before moving his base of operations to Roswell, NM. He also conducted experiments proving rockets work in a vacuum (circa 1912), invented an early form of bazooka (1918), pioneered the use of gyroscopes and jet vanes in rockets (1932), and used film cooling and gimbal mounts (see separate entry) in his designs, among numerous other contributions.

Gravitational parameter
A constant (μ) used in orbital calculations around a given gravitating body, equal either to the product of Newton's universal gravitational constant (G) and the mass (M) of the central body, or to the product of the surface acceleration of gravity (g) and the square of the radius (r) of that body.

$$\mu = GM = gr^2$$

Gravitational acceleration
The change in velocity with respect to time experienced by objects falling in gravity fields, designated by lower case g, and varying from place to place in the universe. On the surface of Earth, the standard acceleration of gravity is

$$g_e = 32.174 \text{ ft/s}^2 = 9.80665 \text{ m/s}^2$$

Gravitational constant
Newton's universal gravitational constant (*G*), not to be confused with the standard acceleration of gravity (g_e) at Earth's surface.

$$G = 6.6726 \times 10^{-11} \text{ N m}^2/\text{kg}^2 = 3.4389 \times 10^{-8} \text{ ft}^3/\text{slug sec}^2$$

Gravity gradient
The magnitude and direction of a local gravitational field.

Gravity loss
The reduction in velocity increment affecting space launch vehicles, due to gravity. Two methods of reducing gravity loss are to accelerate to orbital speed as quickly as possible, or to generate aerodynamic lift in winged vehicles.

Gravity turn
A trajectory followed by a launch vehicle, designed to minimize gravity loss. By gradually turning a launch vehicle sideways during launch, two things happen as it accelerates: (1) Earth's surface begins to fall away from the rocket's flight path due to curvature and (2) centrifugal forces begin to counteract the centripetal forces of gravity.

Gravity well
A geometrical representation of a gravitational field, with the gravitating body at the bottom and the slopes of the sides representing the gravity gradient.

GTOW
Gross takeoff weight.

Gyroscope
An instrument that uses a spinning mass to maintain a fixed orientation in space, attached by a gimbal mount (see separate entry) to its case, and used in attitude and directional control instrumentation and accelerometers.

Heat exchanger
A device consisting of a coiled tube, past which hot turbine exhaust gases (already heated in a gas generator) are passed, in order to transfer heat into a cold liquid propellant or a pressurizing gas carried inside the coil. Heat exchangers can also remove heat from an airflow.

Heat transfer
The conveyance of heat from one body or system to another. Heat may be transferred by conduction, convection, or radiation. Understanding these phenomena is important in keeping a space vehicle at the proper temperature in all flight regimes.

HTHL
Horizontal takeoff horizontal landing, the flight mode of an advanced spaceplane. See also VLHA.

HTPB
Hydroxyl-terminated polybutadiene, a solid rocket fuel also known as rubber.

Hyperbola
An open-ended two-branch curve in which the eccentricity ($e = c/a$) (see separate entry) is greater than 1. Corresponding hyperbolic orbits have specific energies (see separate entry) greater than 0 and result in escape from the central gravitating body.

$$\frac{x^2}{a^2} - \frac{y^2}{b^2} = 1$$

The formula for the hyperbola is similar to that for an *ellipse*. Here, the distance from the origin to each focus is $c = \sqrt{a^2 + b^2}$ measured along the x-axis. Hyperbolas can be drawn by first constructing linear asymptotes with equations $y = bx/a$ and $y = -bx/a$. The curves are then drawn through the points $(a,0)$ and $(-a,0)$, and the foci are located at $(c,0)$ and $(-c,0)$. The two curves will be symmetrical to both the x and y axes, and one of the foci will correspond to the center of a gravitating body, with the curve corresponding to the path of an escaping spacecraft.

Hyperbolic trajectory
An escape orbit followed by an object near a gravitating body. A hyperbolic trajectory has specific energy (see separate entry) greater than 0 and eccentricity (see separate entry) greater than 1. An object on a hyperbolic trajectory relative to a planet may, nevertheless, remain in an elliptical orbit around the Sun.

Hypergolic propellant
A fuel or an oxidizer that spontaneously ignites on contact with the other. Examples are MMH, UDMH, and N_2O_4.

Ideal exhaust velocity
The theoretically maximum velocity of exhaust products from a rocket thrust chamber, given a mean molecular weight of propellants, combustion chamber pressure and temperature, and ambient pressure.

Impulse
The total impulse (I) is the product of a constant thrust (F) and the burn-time (t), or a variable thrust integrated over that time interval. Compare to specific impulse (see separate entry).

$$I = Ft = \int_{t_1}^{t_2} F\,\mathrm{d}t$$

Inertial guidance
A self-contained system by which a space vehicle is controlled by reference to an inertial frame, using a system of three gyroscopes and three accelerometers

mounted on a stabilized gimbal-supported platform. The accelerometers measure the acceleration of the vehicle in three axes, while the gyroscopes provide signals keeping the platform properly aligned. Velocity and position are continuously integrated (calculated) from accelerometer readouts.

Interplanetary travel
Space travel within the solar system.

Interstellar travel
Space travel beyond the solar system.

Jet vane
A movable gyroscopically controlled plane, made of carbon or some other resilient material, placed in the exhaust stream of a rocket nozzle to steer the vehicle.

Kepler's laws
Three laws of planetary motion formulated by Johannes Kepler in the early seventeenth century, using prior detailed observations by Tycho Brahe:

1. A planet moves in an elliptical orbit about the Sun, with the Sun at one focus.
2. A line drawn from the Sun to a planet sweeps out equal areas in equal times.
3. The square of a planet's orbital period is proportional to the cube of its semi-major axis.

LAS
Launch abort system, to be used on the Orion Ares launch vehicles.

Launch vehicle
A rocket used to launch a spacecraft into orbit, typically consisting of two or more liquid or solid propellant stages. In the future, launch vehicles and spacecraft will be integrated into a reusable single-stage-to-orbit vehicle.

L/D
Lift-to-drag ratio, the ratio of aerodynamic lift to atmospheric drag. An airplane with L/D = 20 would provide 20 pounds of lift at the cost of only 1 pound of drag, giving it an excellent mechanical advantage.

LEO
Low-Earth orbit, encompassing altitudes from 100 to 500 miles above the surface, or 200–800 km.

LES
Launch escape system, used on the Apollo Saturn launch vehicles.

Lift
An aerodynamic force caused by a difference in pressure above and below a wing. It is found by the algebraic product of three terms: the coefficient of lift (C_L), the dynamic pressure (q) (see separate entry), and the wing area (S).

$$L = C_L qS = C_L \left(\tfrac{1}{2}\rho v^2\right)S$$

Lifting body
A flight vehicle that develops lift by the shape of its fuselage.

LM
Lunar module, the Apollo two-stage Lunar lander, pronounced "lem."

LSAM
Lunar surface access module, the Orion two-stage Lunar lander.

Mach number
The ratio of vehicle speed to the local speed of sound; this depends upon air temperature and density.

Mass
A measure of the amount of material in a body, expressed in slugs or kilograms. A body's mass stays constant throughout the known universe, in contrast to weight.

Mass flow rate
A quantity used to specify propellant flow rate in a liquid rocket engine, expressed in slugs or kilograms per second, or weight flow rate divided by the standard acceleration of gravity.

$$\dot{m} = \frac{\mathrm{d}m}{\mathrm{d}t} = \frac{\dot{w}}{g_e}$$

Mass ratio
The ratio R of the total initial mass M (or weight W) of a launch vehicle and propellants just before engine ignition, to the final mass m (or weight w) just after end-of-burn cutoff. The higher the mass ratio, the more change in velocity (ΔV) that can be achieved at any specific impulse (see separate entry).

$$R = \frac{M}{m} = \frac{W}{w}$$

Max *q*
Maximum dynamic pressure (q), which is equal to one-half the air density "rho" (ρ) times the velocity squared, or $\frac{1}{2}\rho v^2$. During space launch, dynamic pressure increases exponentially with airspeed while decreasing linearly with air density. Thus, there is typically one point during ascent that experiences "max q."

MMH
Mono-methyl hydrazine, a hypergolic fuel used in many spacecraft attitude control thrusters.

Momentum
The product (p) of the mass (m) and velocity (v) of any object:

$$p = mv$$

Momentum thrust

That portion of a rocket's thrust due to the product of *mass* flow rate $\dot{m}$ and exhaust velocity c. The other component of thrust is called pressure thrust (see separate entry).

$$F_{\mathrm{M}} = \dot{m}c$$

NASA

The US National Aeronautics and Space Administration, organised in 1958 from the former NACA, the National Advisory Committee for Aeronautics.

NASP

The US National Aero-Space Plane, under study in the late 1980s and 1990s.

Newton's laws of motion

Three laws formulated by Sir Isaac Newton during the seventeenth century:

1. A body at rest remains at rest and a body in unaccelerated motion continues in a straight line unless acted upon by an external force.
2. A force (F) imparted to a body gives that body an acceleration (a) inversely proportional to its mass (m). This law may be expressed mathematically by the equation $F = ma$.
3. For every action there is an equal and opposite reaction.

Newton's law of universal gravitation

A mathematical principle discovered by Sir Isaac Newton, stating that every mass (M) attracts every other mass (m) in the universe with a force (F) directly proportional to the product of the two masses and inversely proportional to the square of the distance, or radius, (r) between their centers. The constant of proportionality (G) is Newton's gravitational constant.

$$F = \frac{GMm}{r^2}$$

Oberth, Hermann (1894–1989)

A German rocket pioneer (born in Transylvania) who independently worked out the mathematical details of rocket-powered spaceflight in his 1923 book, *The Rocket to Interplanetary Space*.

OPT

Orbital propellant transfer.

Optimum expansion

The condition that exists when the divergent section of a thrust chamber exactly matches the natural expansion of the exhaust gases at a given ambient pressure. This results in an exhaust stream parallel to the thrust chamber axis, yielding peak efficiency. As a rocket climbs into orbit, its engines typically experience regions of overexpansion, optimum expansion, and underexpansion – in that order – unless they are altitude compensated.

Orbit
The elliptical, parabolic, or hyperbolic path followed by a satellite under the influence of a gravitating body in space.

Oxidizer
An oxygen-containing compound required for combustion of a fuel. Examples are ordinary air, liquid air, liquid oxygen (LOX or LO_2), nitrogen tetroxide (N_2O_4), hydrogen peroxide (H_2O_2), nitrous oxide (N_2O), and ammonium perchlorate (NH_4ClO_4)

Overexpansion
Condition affecting rocket nozzles that are too large for a given ambient pressure at low altitude, typically causing separation of exhaust gases from nozzle walls.

Parabola
An open-ended single curve with an eccentricity of 1, corresponding to a trajectory with specific energy equal to 0. A parabolic orbit would eventually result in the body reaching zero speed at infinite distance.

$$y = ax^2 + k$$

Referring to this standard equation and Cartesian coordinates, the curve is symmetrical to the y-axis and opens upward or downward depending on the sign of a. It will be shifted k units from the x-axis. Interchanging x and y results in a left or right opening parabola.

Payload
That portion of a loaded space vehicle that pays to enter space: the spacecraft atop a launch vehicle or passengers and cargo within a spaceship.

Payload fraction
The fraction of a space launch vehicle, by mass or weight, which is taken up by the payload. The economic goal is to maximize the payload fraction, which conflicts with the engineering requirement to maximize the mass ratio (see separate entry).

Pegasus
A solid rocket that air-launched the X-43 hypersonic scramjet from the belly of a B-52B in 2004.

Performance parameters
Four parameters critical for airplane performance: power loading (weight/power), wing loading (weight/wing surface area), drag coefficient, and maximum lift-to-drag ratio $(L/D)_{max}$.

Perigee
The lowest point in an elliptical orbit around Earth.

Pogo effect
A longitudinal oscillation in a rocket induced by propellant flow instabilities.

Pressure
A force per unit area, expressed in pounds per square inch (psi) or pounds per square foot (psf). In the SI system, pressure is expressed in newtons per square meter, called pascals.

Pressure thrust
That portion of a rocket's thrust due to the difference between the ambient pressure (P_A) and nozzle exit pressure (P_E) acting on the nozzle exit area (A_E), the other portion being the momentum thrust (see separate entry).

$$F_P = (P_E - P_A)A_E$$

Propellant
A combustible substance, a fuel or an oxidizer, burned in a rocket engine and used to propel a space vehicle.

Ramjet
A simple air-breathing jet engine, with no moving parts, that utilizes ram air pressure in place of a turbine and compressor to sustain combustion and produce thrust.

RCS
Reaction control system.

Regenerative cooling
A method of cooling a rocket engine, in which one of the propellants is circulated through a cooling jacket in the thrust chamber (see separate entry) before being introduced to the combustion chamber (see separate entry).

Rocket
A device that accelerates by means of the conservation of momentum and Newton's third law. The momentum of the exhaust gases in one direction equals the momentum of the rocket in the other direction.

Rocket equation
The fundamental equation of rocket flight, relating the effective exhaust velocity (c) of the expelled gases and the mass ratio (M/m) (see separate entry) of the rocket to the achievable velocity increment (ΔV):

$$\Delta V = c \ln(R) = I_{sp} g_e \ln(M/m)$$

In exponential form this becomes

$$R = e^{\Delta V / c}$$

Sänger, Eugen (1905–1964)
An Austrian-German rocket engineer, designer of the *Silbervogel* antipodal bomber, and author of *Raketenflugtechnik* (*Rocketplane Engineering*).

Satellite
Any body that orbits another, usually more massive, body.

Saturn V
The eight-piece 363-ft Apollo launch vehicle and Moonship, consisting of three main rocket stages, the two-piece Lunar Module, the combined Command and Service Modules, and the Launch Escape System. The five F-1 engines of the first stage produced a total of 7½ million pounds of thrust to lift the 6½ million pound vehicle off the launch pad. The only piece that returned to Earth was the three-man conical Command Module.

Scramjet
Supersonic combustion ramjet, in which combustion is maintained in a supersonic airflow within the engine. Unlike ramjets, scramjets have no diffuser and are typically integrated with the airframe.

Shock diamond
A visible shock wave in a rocket exhaust stream, which may appear as a row of luminous stationary spots.

Shock wave
A thin region where pressure, density, and temperature drastically change as a result of supersonic flow. It may also be produced by an explosion, lightning, or rocket engine.

Slug
A unit of mass in the US customary system of measurement, which has a weight of 32.174 pounds at the surface of Earth.

$$1 \text{ slug} = 1 \text{lb s}^2/\text{ft} = 14.59 \text{ kg}$$

Slush hydrogen
A mixture of liquid and solid hydrogen, with a 16% higher density and 18% higher heat capacity than liquid hydrogen. Its main advantage is to reduce the gross takeoff weight of any hydrogen fueled spacecraft because of smaller fuel tank requirements. For this reason, it was selected for the NASP project in the mid-1990s.

Specific energy
The sum of the kinetic and potential energies per unit mass of a space vehicle, expressed as

$$E = \frac{v^2}{2} - \frac{\mu}{r}$$

where v is the velocity of the spacecraft relative to a gravitating body, μ is the gravitational parameter (see separate entry) for the central body in question, and r is the distance of the vehicle from the center of the gravitating body.

Specific impulse
The ratio of thrust to fuel consumption, or the time in seconds that one pound of onboard propellant will provide one pound of thrust. This can be expressed as the thrust force (F) in pounds divided by the propellant weight flow rate ($\dot{w} = \dot{m}g_e$) in pounds per second; as the total impulse $I = Ft$ divided by the total weight (w) of propellant consumed; or as the ratio of effective exhaust velocity (c) (see separate entry) to the standard acceleration of gravity (g_e) on Earth.

$$I_{sp} = \frac{F}{\dot{w}} = \frac{F}{\dot{m}g_e} = \frac{Ft}{w} = \frac{I}{w} = \frac{c}{g_e}$$

SRB
Solid rocket booster.

SSTO
Single-stage-to-orbit.

Stage
A section of a rocket that drops off when its propellants are expended, thereby lightening the load and increasing the overall mass ratio (see separate entry) of the launch vehicle.

Strake
An aerodynamically shaped component or compartment, mounted along a spaceplane fuselage, often used for carrying extra propellants.

Suborbit
A ballistic trajectory followed by a space vehicle possessing insufficient velocity or energy to attain orbit.

Thrust
A force (F) delivered by a rocket thrust chamber, expressed in pounds or newtons, and resulting in an acceleration of a spacecraft or launch vehicle in accordance with Sir Isaac Newton's laws of motion. Thrust force comprises the sum of a momentum thrust and a pressure thrust component:

$$F = \dot{m}c + (P_E - P_A)A_E$$

where $\dot{m}$ is propellant mass flow rate (slugs or kg/s), c is exhaust velocity (ft or m/s), P_E is nozzle exit pressure (psi or pascals), P_A is ambient pressure (psi or pascals), and A_E is nozzle exit area (in.^{2} or m^2). Because P_A decreases with altitude, thrust always reaches a peak value in vacuum conditions.

Thrust cell
One of many small thrust chambers (see separate entry) of low area ratio (see separate entry) in an aerospike engine; these emit their exhaust products onto an annular or linear ramp.

Thrust chamber
The combustion chamber (see separate entry), throat, and nozzle of a conventional rocket engine.

Thrust mount
A typically steel tubular structure designed to transmit thrust forces from the rocket engine to the space vehicle itself.

Thrust vector control
Means by which a gimbaled thrust chamber can vector its thrust axis, thereby controlling the direction of flight.

TLI
Trans-lunar insertion, the maneuver used to insert a spacecraft on a trajectory for the Moon.

TPS
Thermal protection system.

Trajectory
A path through space governed by the laws of Sir Isaac Newton.

Tsiolkovskiy, Konstantin E. (1857–1935)
Russian rocket scientist, the "father of cosmonautics." A self-educated mathematics and physics teacher, he was the first to work out the theoretical details of rocket-powered spaceflight, including the use of liquid propellants, jet vanes, and staging.

TSTO
Two-stage-to-orbit.

Turbopump
A turbine-driven pump, which in rockets is used to deliver liquid propellants to the combustion chamber.

UDMH
Unsymmetrical di-methyl hydrazine, a toxic and corrosive hypergolic fuel used in the reaction control systems of many spacecraft.

Ullage
The intentional empty space in a liquid propellant tank.

Underexpansion
A condition affecting rocket nozzles that are too small for a given ambient pressure at high altitude, typically causing ballooning of exhaust gases behind the nozzle.

Vector
A physical quantity having both magnitude and direction, represented graphically as an arrow. Examples include velocity, thrust, drag, and acceleration.

Velocity
A vector quantity specifying both the speed and direction of motion.

Velocity increment
The theoretical delta-V (ΔV), or change in velocity, either gained or lost by a space vehicle due to a rocket impulse. The actual ΔV may be less than this, due to gravity, drag, and other losses. Delta-V is directly proportional to exhaust velocity (c) and logarithmically proportional to *mass ratio* (M/m).

$$\Delta V = c \ln\left(\frac{M}{m}\right)$$

Vernier engines
Small rocket engines used to fine-tune the final velocity of a launch vehicle.

Viscous interaction
A phenomenon whereby a thickening boundary layer may interact with an inviscid flowfield at supersonic speeds. As high-speed heating occurs, the boundary layer expands due to increased viscosity and decreased density, causing it to interact with inviscid regions farther from the vehicle and severely increasing drag. Lift, stability, and heat transfer are also affected by viscous interaction.

VLHA
Vertical landing in a horizontal attitude, also vertical lift-off in a horizontal attitude, describing Harrier-like operations for ultra-advanced spaceplanes. This flight mode operates within HTHL.

VTHL
Vertical takeoff horizontal landing, as in the Space Shuttle.

VTVL
Vertical takeoff vertical landing. Compare *HTHL* and *VLHA*.

Weight
The force of gravity that pulls on a mass, different on every gravitating body in the universe.

Wing loading
The pressure borne by a wing, determined by dividing gross vehicle weight by the wing area. Airplanes experience typical wing loadings of anywhere from 15 (Cessna 172) to 100 (C-5 Galaxy) pounds per square foot.

Working mass
The fluid material that is accelerated through and exhausted from a jet or rocket engine to provide propulsion.

Von Braun, Wernher (1912–1977)
A German-American rocket scientist, motivated by the quest for manned spaceflight. An early member of the German Society for Spaceship Travel, he designed the V-2 rocket and later helped NASA develop its Saturn V launch vehicle for the Apollo program.

Appendix B
Spaceplane Mathematics

This section is arranged alphabetically for ease of use. It is not meant to be read from beginning to end, but section by section. For the novice, a good place to begin might be the Greek alphabet, followed by units and then conversions.

Centrifugal Force

Many physicists claim that centrifugal force does not exist, only centripetal force, which pulls the body toward the center of revolution. Test pilots might disagree. In any case, it is the centrifugal force balancing with the gravitational force that keeps an orbiting body in continuous free-fall.

Centrifugal force = mass × velocity squared divided by radius

$$F_c = mv^2 / r$$

To understand the units in this equation, read the *slug* section below. When you multiply all the units out, centrifugal force should fall out of the equation in pounds.

Example: Two spaceplanes are connected by a 100 ft collapsible dorsal tunnel. At what speed must they rotate about their common center of gravity to generate artificial Mars gravity onboard each spaceplane? Mars surface gravity is 37% that at the surface of Earth.

Solution: Realize, from Newton's Second Law, that centrifugal force is really centrifugal acceleration x mass. Secondly, realize that centrifugal acceleration is equivalent to gravitational acceleration, which on Earth is 32.174 ft/s^2.

$F_c = ma_c = mv^2 /r$
$a_c = v^2 /r$
$0.37g_e = v^2 /50\text{ft}$
$\text{v} = (50 \text{ ft} \times 0.37 \times 32.174 \text{ ft/s}^2)^{1/2}$
$\text{v} = 24.397$ ft/s Speed at rim

Circumference of rotation = C = $2\pi r$ = 2π(50 ft) = 314.16 ft
Rotation rate = Speed at rim (ft/s)/Circumference of rotation (ft/rev) = rev/s
Rotation rate = (24.397 ft/s)/(314.16 ft/rev) = 0.077658 rev/s
Revolution period (sec/rev) = 1/Rotation rate = 1/0.077658 rev/sec = 12.877 sec/rev

With this information, the crew may want to either (a) lengthen the distance between their craft or (b) slow down the rotation rate and settle for a lower artificial g-field.

Conversions

It is always useful to know how to convert from one system of units into another, and with a little practice it becomes easy. All you need is a basic hand calculator. Here are some exact conversion factors and a few examples. It might behoove the reader to read the section on units first.

1in. = 2.54 cm
1mile = 1.609344 km
1ft = 0.3048 m

Here is how to use these conversion factors. Since these are equalities, one side of the equation can be divided by the other side, and you will still end up with an answer of one, because anything divided by itself is always unity. The conversion factors are 2.54 cm/in., 1.609344 km/mile, and 0.3048 m/ft.

Example 1: Convert 7,956 miles into kilometers:
7,956 miles × 1.609344 km/mile = 12,804 km
This is the diameter of Earth.

Example 2: Convert 299,792 km into miles:
299,792 km/(1.609344 km/mile) = 186,282 miles
This is the approximate distance light travels every second.

Example 3: Now do the first problem using feet and meters:
[1mile = 5,280 ft and 1km = 1,000 m]
7,956 miles × (5,280 ft/mile) × (0.3048 m/ft)/(1,000 m/km)
42,007,680 ft × 0.0003048 km/ft
12,804 km

Same answer. How about that!

How do you know when to multiply the conversion factor, and when to divide? Simple. Just look at the units. Always make sure the units you are trying to convert from cancel out, and the units you are trying to end up with are in the numerator. Then you will always get the right answer.

Dynamic Pressure

Dynamic pressure = 1/2 air density × velocity squared,

$$Q = \frac{1}{2}\rho v^2$$

This equation shows at a glance that dynamic pressure goes down linearly (straight line on a pressure vs. time graph) and goes up exponentially (steepening curved line) as a rocket accelerates and climbs into rarefied air. Air density (ρ) is the linear factor, while velocity (v) is the exponential factor.

Exhaust Velocity

Exhaust velocity = specific impulse × Earth standard acceleration of gravity:

$$c = I_{sp}\, g_e$$

This equation works anywhere, even in space, using Earth standard gravitational acceleration, because specific impulse is defined as pounds thrust divided by propellant "weight" (Earth pounds) flow rate. Dimensional analysis shows how this works:

$$c = [\text{lb} / (\text{lb/s})][\text{ft/s}^2]$$
$$c = [\text{s}][\text{ft/s}^2]$$
$$c = \text{ft/s}$$

Greek Alphabet

What are all those funny looking letters in the equations? If they are Greek to you, then you are right, because they are Greek letters. Here is the list.

Α	α	alpha	Ι	ι	iota	Ρ	ρ	rho
Β	β	beta	Κ	κ	kappa	Σ	σ	sigma
Γ	γ	gamma	Λ	λ	lambda	Τ	τ	tau
Δ	δ	delta	Μ	μ	mu	Υ	υ	upsilon
Ε	ε	epsilon	Ν	ν	nu	Φ	φ	phi
Ζ	ζ	zeta	Ξ	ξ	xi	Χ	χ	chi
Η	η	eta	Ο	ο	omicron	Ψ	ψ	psi
Θ	θ	theta	Π	π	pi	Ω	ω	omega

Mass and Weight

Mass is simply the total amount of material in an object. It can be thought of as a tally of all the subatomic particles – electrons, protons, and neutrons – in a body. With this in mind, it is easy to see that mass remains constant everywhere in the universe. Mass is expressed in kilograms or slugs.

Weight is the force between a mass and an accelerating surface or gravitating body. It depends on three things: an object's mass, the strength of the local acceleration field, and the state of motion of the object relative to the field. The weight of an object is just its mass times the acceleration of gravity acting on that object. It is expressed in newtons or pounds.

Example 1: Find the universal mass of Astrid the Swedish astronaut, in slugs, who weighs 21 lbs at a Moonbase, before she dons her spacesuit. The acceleration of gravity on Earth is 32.17 ft/s^2 and the acceleration of Lunar gravity is 1/6 that on Earth.

Solution: Using Newton's second law, we know that weight equals mass times acceleration of gravity:

$$W = mg$$

Solving for mass, we have

$$m = W/g$$

Plugging in numbers:

$$m = (21 \text{ lb} \times 6) / (32.17 \text{ ft/s}^2)$$
$$m = 126 \text{ lb}/ (32.17 \text{ ft/s}^2)$$
$$m = 3.9 \text{ slugs}$$

Examining the problem, we see Astrid's 126-lb Earth weight appear in the numerator. The astronaut has a universal mass of 3.9 slugs, an Earth weight of 126 lb, and a Lunar weight of 21 lb.

Example 2: If 1 kg weighs 2.2 lb on Earth, what is the Astrid's mass in kilograms and weight in newtons on Moon and Earth? The acceleration of gravity on Earth is 9.807 m/s^2.

Solution:

$$(126 \text{ lb}) / (2.2 \text{ lb/kg}) = 57\text{kg}$$
$$W_E = mg = (57\text{kg})(9.807\text{m/s}^2) = 560\text{N}$$
$$W_M = (57\text{kg})(1/6)(9.807\text{m/s}^2) = 93\text{N}$$

The astronaut has a universal mass of 57 kg, an Earth weight of 560 N, and a Lunar weight of 93 N.

Per

Whenever you see a slash (/) in numerical calculations, it signifies "per" or "divided by."

For example, the standard acceleration of gravity at the surface of planet Earth is 32.174 feet per second squared. Expressed mathematically, this information is given simply as $g_e = 32.174$ ft/s^2.

Rocket Equation

Delta V = exhaust velocity × natural log of mass ratio

$\Delta V = c \ln (M/m)$

Significant Figures

When doing mathematical calculations, it is important to employ the correct number of digits in the calculation, and finally, to express the correct number of significant figures in the result. Generally, the number of significant figures in the answer is no more than the number of significant figures in any of the values you used in the calculation.

Convert one million kilometers into miles:

1,000,000 km / 1.609344 km/mile = 621,371.192237 miles = 600,000 miles

Convert 238,000 miles into kilometers:

238,000 miles × 1.609344 km/mile = 383,023.872 km = 383,000 km

Slug

This is the unit of mass in the US customary system of measurement, which has a weight of 32.174 lb at the surface of Earth.

$$1 \text{ slug} = 1 \text{ lb sec}^2 / \text{ft} = 14.59 \text{ kg}$$

Specific Impulse

One of the most important parameters in rocket engine performance, specific impulse is given in units of seconds. It is the number of seconds that 1 lb of propellant will provide 1 lb of thrust.

Specific impulse = thrust force divided by propellant "weight" flow rate

$$I_{sp} = F/\dot{w}*$$

If the atmosphere is providing the working mass, then specific impulse goes up significantly, because the propellant weight flow rate refers only to onboard propellants, not to exterior derived working mass.

The propellant "weight" flow rate is actually the mass flow rate times the standard acceleration of gravity at the surface of Earth.

Units

Whenever a number is expressed, a unit of some kind is used to specify what the number refers to. These are the little words that come after the number, and they help us decide whether to multiply or divide when we do conversions. Units must always be applied consistently and logically. If a Russian cosmonaut is 2 m tall, and an American astronaut is 6 ft tall, the difference in their heights is not 4 m-ft, even though 6 – 2 = 4. The actual difference must be expressed in a single unit.

[2 m] – [6 ft]
[(2m) / (0.3048 m/ft)] – [6 ft]
[6.56167979003 ft] – [6 ft]
0.56167979003 ft
6.74 in.

Vectors

Think of a vector as an arrow. It has both direction (the way it is pointed) and magnitude (the length of the arrow). Velocity is one example of a vector, as are acceleration and all flight forces (lift, weight, thrust, and drag).

Vehicle Volume

When the size of a three-dimensional structure is doubled, the volume increases by 2^3, or eight times, but the weight merely doubles. When its size is tripled, volume goes up by 3^3 or 27 times. Again, the weight only triples. A modest increase in size therefore results in a huge increase in volume, without a concurrent weight penalty.

Assume the propellant-carrying capacity of a spaceplane is represented by two volumes, a right circular cylinder and two hollow triangular wings. The two triangular wings can be mathematically combined to create a single rectangular volume that can be calculated easily. The formula will have two terms, one for the cylinder and one for the wings. We will need the radius and length of the fuselage, as well as the length, width, and height of the wing.

$$V = l\pi r^2 + lwh$$

This simplified formula reveals that if the dimensions (length l, width w, height h, and radius r) are increased, the volume will once again go up by the cube of that increase. The first term guarantees this result because length and the square of the radius are multiplied together, and the second term ensures this result because length, width, and height are all multiplied together.

The unmistakable conclusion is that if a vehicle, say, four times the size of the X-15 had been constructed, it would have had a propellant capacity of 4 cubed, or 64 times as much. Of course, the fully fueled vehicle would also have weighed 64 times as much, making it 64 times more difficult to accelerate into space.

Appendix C
Spaceplane Timeline

Appendix C

1903 (December 17)	Wright brothers make the first four powered heavier-than-air flights
1919	Robert Goddard publishes "A Method of Reaching Extreme Altitudes"
1926 (March 16)	Robert Goddard flies the first liquid propellant rocket, Auburn, MA
1928	Fritz Stamer flies 1.2 km in catapult-launched black powder rocketplane *Ente*
1929	Yuri Kondratyuk proposes lifting body to decelerate through atmosphere
1929	Fritz von Opel flies RAK.1 solid-propellant rocket glider near Frankfurt, Germany
1930	Frank Whittle gets a patent for the jet engine
	Gottlob Espenlaub reaches 90 km/h in rocket glider, Düsseldorf, Germany
1931	Dr. Eugen Sänger begins liquid fuel rocket experiments, Vienna, Austria
1931	Reinhold Tiling flies aluminum solid fuel rocket with folding wings
1931 (June 9)	Robert Goddard receives patent for "rocket-propelled aircraft"
1933	Eugen Sänger publishes *Raketenflugtechnik* (*Rocket Flight Engineering*)
1934	William Swan ascends to 200 ft in solid-fuel rocketplane, Atlantic City, NJ

1937	H_2O_2 Walter engine assists takeoff of Heinkel He-72
	1,000 kg_f thrust liquid fuel rocket is flight-tested in Heinkel He-112
1938	Eugen Sänger and Irene Bredt begin research on antipodal rocket bomber
1939	Heinkel He-176 rocketplane begins test flights at Peenemünde
1940	Tow-launched variable thrust Korolev RP-318-1 glide-rocket reaches 200 km/h
1941	Messerschmitt Me 163A sets world speed record of 1004.5 km/h
1944	Me 163 becomes first operational rocketplane
1945	Winged V-2 demonstrates Mach 4 glide
1947 (October 14)	Chuck Yeager in Bell X-1 flies faster than sound, reaching Mach 1.06
1947 (November 6)	X-1 reaches Mach 1.35, roughly 30 times the speed of Wright's Flyer
1951	Douglas test pilot Bill Bridgeman takes D-558-2 Skyrocket to record 79,500 ft.
1953	First level-flight supersonic jet aircraft, the F-100 Super Sabre, makes maiden flight
1953	NACA test pilot Scott Crossfield takes Douglas Skyrocket to Mach 2.
1956	Iven Kincheloe in Bell X-2 reaches 126,200 ft, first pilot to exceed 100,000 ft
	Mel Apt in Bell X-2 loses control, is killed, after being first to reach Mach 3
	Two F-100 Super Sabres perform the first successful "buddy" refueling
1961	X-15 makes first aircraft flight exceeding Mach 4, then 5, then 6
	X-15 makes first aircraft flight above 200,000 ft
1962	X-15 makes first flight exceeding dynamic pressure of 2,000 psf
	X-15 makes first aircraft flight above 300,000 ft and first above 50 miles
1963	X-15 makes first aircraft flight above 100 km, reaching 347,800 ft.
	X-15 makes second aircraft flight above 100 km, reaching 354,200 ft

1966	X-15 experiences its highest dynamic pressure of 2,202 psf
	X-15 sets world absolute speed record of Mach 6.33 or 4,250 mph
1966	First flight of the Northrop HL-10 lifting body
1967	X-15 reaches Mach 6.70 or 4,520 mph – unofficial world speed record
	X-15 pilot-astronaut Michael J. Adams is killed by severe pitch oscillations during reentry
1969	X-24A makes its first drop-glide flight
	Northrup HL-10 becomes first lifting body to fly faster than sound
1970	HL-10 achieves fastest flight of any lifting body, Mach 1.86 or 1,228 mph
1972	Bill Dana takes M2-F3 lifting body to its top speed of 1,064 mph or Mach 1.6
	John Manke flies M2-F3 lifting body to its highest altitude of 71,500 ft
1973	X-24B lifting body makes first drop-glide flight, Edwards AFB, California
1974	X-24B reaches its maximum speed of 1,164 mph
1975	X-24B reaches its maximum altitude of 74,130 ft
1977 (August 12)	Space Shuttle *Enterprise* makes first glide-flight from 747 carrier, Edwards AFB, California
1981 (April 12)	Space Shuttle *Columbia* makes maiden voyage – first flight of an orbital spaceplane
1986 (January 28)	Space Shuttle *Challenger* breaks up 73 s after launch, killing her crew of 7
1988 (November 15)	Soviet Spaceplane *Buran* makes its only spaceflight, unmanned, with automatic landing.
2001	First X-43 test flight fails when Pegasus booster loses control
2002 (May 12)	*Buran* spaceplane is destroyed when hangar collapses
2003 (February 1)	Space Shuttle *Columbia* breaks up on reentry, taking her crew of 7
(December 17)	SpaceShipOne makes its first powered flight 100 years after Wright brothers
2004	X-43 unmanned scramjet sets world airspeed record of Mach 6.83
	Mojave Airport, California, becomes Mojave Spaceport.

(June 21)	SpaceShipOne makes first private spaceflight, reaches 100.1 km
(September 29)	SpaceShipOne reaches 102.9 km
(October 4)	SpaceShipOne reaches 112.0 km and wins the X-prize
	X-43 unmanned scramjet sets new airspeed record of Mach 9.68
2006	Genesis 1, first inflatable subscale space module, enters orbit on Dnepr booster
2007	X-48B unmanned blended wing body subscale test vehicle makes first flight

Appendix D
Spaceplane Data

The purpose of this section is to list, categorize, and detail specific data on several spaceplane or spaceplane-related vehicles.

The first section will outline all of the known spaceplanes, categorized according to whether they are (I) atmospheric, (II) suborbital, or (III) orbital. The orbital spaceplanes are further classified as to whether they are (A) ballistically launched, (B) TSTO/piggyback designs, or (C) single-stage-to-orbit SSTO.

The second section will be an alphabetical listing of spaceplane models with details of each.

Spaceplanes Classified

I. Atmospheric Rocketplanes/Lifting Bodies/Test Vehicles

A. Aircraft-Launched
Bell X-1 (1947) Mach 1 1947 Chuck Yeager
Douglas Skyrocket D-558–2 (1948–1956) Mach 2 1953 Scott Crossfield
Bell X-2 (1952–1956) Mach 3 1956 Mel Apt
M2-F1
M2-F2
M2-F3
HL-10
Martin Marietta X-24A (1969–1971)
X-24B
Martin Marietta X-24C (1973–1975) X-24C (1976)
Boeing X-40 (1998–2001)
NASA X-38 (1999–2002)
Orbital Sciences X-34 (canceled 2001)
EADS Phoenix (2004)
Boeing X-37 (2006)

B. Self-Launched
Me 163 Komet (1944)
MiG-105 Spiral (1976–1978)

OK-GLI Buran Aerodynamic Analog (1985–1988)
Rotary Rocket Roton (1999)
XCOR EZ-Rocket (2001)
XCOR X-Racer (2008)

II. Suborbital Spaceplanes

A. Aircraft-Launched
North American X-15 (1959–1968)
SpaceShipOne (2004)
SpaceShipTwo
Space Adventures C-21
Space Adventures Explorer

B. Runway-Launched
Ascender (2000)
Rocketplane XP (planned)
Spacefleet SF-01 (2008)
XCOR Xerus (2008)

C. Ballistic
McDonnell ASSET (1963–1965)
Martin Marietta X-23 PRIME (1966–1967)
Lockheed-Martin X-33 (circa 1999)

III. Orbital Spaceplanes

A. Staged Ballistic
Dyna-Soar (1957–1963)
Kliper (2005)
HOPE-X
MUSTARD
Orbital Space Plane (4 in 1)
USAF X-37B Orbital Test Vehicle (2008)
NASA Space Shuttle (1981–2010)
Hermes (1987–1993)
Buran (1988) canceled 1993
SpaceDev Dream Chaser (2008)
Silver Dart (2009)

B. SSTO Ballistic
VentureStar

B. Aircraft-Launched HTHL TSTO
Blackstar
Sänger II

C. Runway or Track-Launched HTHL SSTO
Silbervogel (1935–1945)
HOTOL (1986–1988)

X-30 NASP (1990–1993) Rockwell International
Avatar RLV (India)
Skylon (planned)
Hopper (planned)

D. HTHL with Aerial Propellant Transfer
Black Horse

Spaceplane Models

Ascender – The second stage of a TSTO spaceplane design conceived by David Ashford of Bristol Spaceplanes, England. It would ride piggyback on a runway-launched booster before entering orbit by itself. Ascender could also take off from a runway using jet engines and climb to 100 km using rocket engines and onboard propellants for suborbital space tourism flights. *Conceptual.*

Black Horse – USAF one-person spaceplane that would take off under its own rocket power, rendezvous with a conventional aerial tanker, take on its full hydrogen peroxide (H_2O_2) oxidizer supply, and fly into orbit from some 40,000 ft. Required good subsonic lift-to-drag ratio. *Conceptual mid-1990s.*

Blackstar – USAF air-launched manned spaceplane with aerospike engines, released from large carrier aircraft at 100,000 ft for surprise orbital reconnaissance. *Rumored 2006.*

BoMi – Bell Aircraft two-stage fully reusable suborbital bomber-missile or reconnaissance vehicle. Both manned delta-wing booster and manned double-delta glide-rocket would land horizontally. Booster: 120 ft long, 60 ft wingspan, 2 crew. Glide rocket: 60 ft long, 35 ft span, 1 crew. Propellants: UDMH and N_2O_4. *Conceptual 1952–1955.*

BOR-4 – Unmanned Soviet orbital test vehicle to assess thermal protection system for Buran. The lifting body design was a downscaled unmanned version of the manned Spiral spaceplane. It was used to evaluate the hypersonic entry and cross range capability of the design, as well as to test heat shield materials for the Buran program. It has excellent characteristics in this and in subsonic flight and resulted in Langley Research center using the shape as the basis for its own HL-20 Crew Escape Rescue Vehicle or Personnel Launch System. After a circuit of the Earth, the BOR-4 spacecraft would deorbit, perform a gliding reentry, follwed by parachute deployment, splashdown in the ocean, and be recovered by Soviet naval forces. *Flew unmanned 5x 1980-1984.*

BOR-5 – Unmanned Soviet suborbital test vehicle using 1/8 scale 2,800-pound model to assess aerodynamic and thermodynamic conditions of Buran shape. *Flew 5 times 1983–1988.*

Brass Bell – A USAF-funded program at Bell Aircraft to develop a two-stage manned suborbital reconnaissance vehicle based on Atlas missile engines. *Conceptual 1956.*

Buran – Soviet space shuttle; made one unmanned flight on November 15, 1988, with automatic landing at Baikonur Cosmodrome. Main differences from U.S. Space Shuttle were, main engines were not carried on the winged orbiter but at base of the Energiya rocket; strap-on rockets were powered by kerosene and LOX instead of solid propellants; and launch vehicle was assembled horizontally and erected before launch. *Flew unmanned 1× 1988.*

DC-X Delta Clipper – A McDonnell Douglas VTVL spacecraft built and flown in the early 1990s. Despite many successful atmospheric test flights, the unmanned vehicle is best remembered on its final flight for tipping over and exploding when one of its four landing legs failed to extend. *Flew in 1990s.*

Douglas Skyrocket – Officially the D-558-2, the first manned rocketplane to reach Mach 2, on November 20, 1953, piloted by Scott Crossfield. *Flew 1948–1956.*

EADS Astrium Space Tourism Project – a concept to take off from a runway with a small jet-powered rocketplane, fly up to 12 km using jet engines only, then accelerate almost straight up using rocket engines fueled by methane and liquid oxygen. A pilot and four passengers would reach 100 km altitude in this manner, the edge of space. *Currently under development.*

EADS Phoenix – A winged testbed, successor to Hermes and predecessor to Hopper, it was a one-sixth scale prototype with a 6.9 m length, 3.9 m span, and 1,200 kg mass. On May 8, 2004, the testbed was dropped from 2.4 km by helicopter and made a GPS-guided 90 s glide to the European Space Range in Kiruna, Sweden, 1,240 km north of Stockholm.

EZ-Rocket – XCOR Aerospace's rocketplane, using a Rutan Long-EZ airframe with its engine replaced by an XCOR in-house liquid rocket engine. The EZ-Rocket has made 26 test flights, including numerous in-flight restarts of the rocket engine. *Flew 26× 2000–2005.*

Hermes – European manned spaceplane project, to be launched by Ariane 5 rocket for crew and cargo flights to the Columbus space station. It originally consisted of a spaceplane with a capacity of six astronauts and 4,500 kg of cargo, but this was reduced to three astronauts with ejection seats after the *Challenger* tragedy in 1986. Hermes was to be equipped with a rear-mounted Resource Module, which would be jettisoned before re-entry. *Conceptual 1987–1993.*

HL-10 – The Northrup Horizontal Lander was a NASA delta planiform lifting body, used to assess the flight and landing capabilities of the lifting body shape, and supported development of the Space Shuttle. *Flew 37× 1966–1970.*

HOPE – Japanese H-II orbital plane, a ballistic orbital spaceplane, to be launched by the H-II rocket, intended to deliver cargo to a space station. *Conceptual 1987–2003.*

Hopper – ESA concept to launch a spaceplane from a 4-km magnetic track, greatly reducing propellant requirements for entering space. *Conceptual.*

HOTOL – A single-stage-to-orbit HTHL spaceplane under development by the British government from 1991 to 1995. *Conceptual.*

Kliper – Russian lifting body spaceplane designed to be ballistically launched and reenter at an angle to reduce G forces on crew. *Conceptual 2004 to present.*

M2-F1 – Flying Bathtub. *Flew > 477× 1963–1965.*

M2-F2 – Initial design had only two tail fins, later modified into the M2-F3. *Flew 1966–1971.*

M2-F3 – Modified from M2-F2 by addition of tall center fin. *Flew 1970–1972.*

Me 163 Komet – World's first operational rocketplane. Flown by Germany in World War II as an interceptor-attack fighter. Endurance 8–12 min. *Flew 1944–1945.*

Orbital Space Plane – Four design concepts intended to replace the Space Shuttle program involving either (1) Apollo-type capsule, (2) lifting body, (3) winged slender-body, or (4) winged fat-body designs. The capsule was ultimately chosen as the *Orion* replacement for the Space Shuttle.

Rocketplane XP – Runway-launched sub-orbital spaceplane being developed by Rocketplane Global of Burns Flat, Oklahoma, for the space tourism industry. *Under active development.*

Roton – Rotary Rocket's unique spaceplane design, which was actually a space helicopter. The rotors used small tip-mounted rocket engines to generate initial lift at takeoff, followed by main rocket engine ignition. For landing, the rotors would autorotate to provide a soft touchdown. The main engine sported a rotating aerospike for automatic altitude compensation.

Sänger II – German project of the 1980s to resurrect Eugen Sänger's two-stage-to-orbit design. The piggyback spaceplane would take off from a runway, separate at altitude, and fly into space. *Conceptual 1980s.*

Silbervogel – "Silverbird" was Eugen Sänger's original design for a manned suborbital spaceplane originally conceived as an "Amerika Bomber," able to deliver a bomb half-way around the world, hence the name antipodal bomber. *Conceptual 1940s.*

SF-01 Spacefleet Project – Private British collaboration using clean propellants (LOX and LH_2) and eight engines to transport up to eight space tourists and two pilots at a time up to a maximum apogee of 340 km, well above space height of 100 km. This would provide the advantage of longer periods of weightlessness and better views of Earth below. *Conceptual.*

Skylon – An improved version of the British HOTOL spaceplane, designed by Alan Bond of Reaction Engines Limited. Taking off horizontally, the unmanned vehicle would employ liquid hydrogen in a novel engine design utilizing a closed helium loop and super-cooled air before entering orbit at the top of the atmosphere. *Conceptual.*

Space Adventures Explorer – Suborbital aircraft spaceplane to be operated from the Ras Al Khaimah spaceport in the United Arab Emirates. It is based on the Space Adventures C-21, designed by the Russian Myasishchev Design Bureau.

SpaceShipOne – The first civilian spaceship and spaceplane, winner of the X Prize, exceeded 100 km twice in 2 weeks in October 2004, from the Mojave Spaceport, California, using hybrid rocket engine with liquid oxidizer, and folding tail. *Flew 2000–2004.*

SpaceShipTwo – Follow-up to SS1, designed to carry two pilots and five passenger-astronauts on suborbital flights beginning about 2009. *Under construction.*

SpaceShipThree – Orbital successor to SS1 and SS2, able to take space tourists to an orbiting space hotel or space station. Method of launch has not yet been revealed. *Conceptual as of 2007.*

Space Shuttle – American spaceplane, first flown on April 12, 1981, consisting of fully reusable double-delta winged orbiter, reusable solid propellant strap-on boosters, and throwaway external tank (ET) filled with liquid hydrogen and oxygen. Typically carries crew of seven and 55,000 pound payload to LEO. Responsible for delivering most International Space Station modules, repairing the Hubble Space Telescope, and conducting numerous satellite and planetary spacecraft delivery and scientific missions. *Challenger* lost on January 28, 1986, due to SRB joint seal failure and subsequent detonation of ET. *Columbia* lost on February 1, 2003, due to left wing damage from foam strike on liftoff, followed by wing failure and vehicle breakup on re-entry. *Enterprise* used for approach and landing tests, 1977, never flew in space. *Discovery*, *Atlantis*, and *Endeavour* still on flight status as of late 2007. Shuttle fleet due for retirement in 2010. *Flew 120× as of late 2007.*

SR-71 Blackbird – High-altitude Mach 3 reconnaissance aircraft, the fastest turbojet powered aircraft known. *Operational.*

VentureStar – Lockheed-Martin orbital single-stage VTHL spaceplane and successor to the X-33 proof-of-concept vehicle. The spacecraft was a fully reusable combined launch vehicle, orbiter, and spaceplane using linear aerospike engines for greatly increased efficiency. It sported a lifting body design with small wings for control. Studied mid-1990s. *Conceptual.*

Winged V-2 – V-2 fitted with wings for added range. Achieved Mach 4 in January 1945. *Flew 2× 1945.*

XCOR Xerus – Suborbital runway-launched spaceplane under development by XCOR Aerospace. It uses in-house piston–pump liquid propellant engines and burns methane or kerosene and liquid oxygen. *Under development 2007.*

X-20 Dyna-Soar – One-person ballistic orbital reconnaissance spaceplane or orbital bombardment vehicle to be launched by Titan III liquid rocket with solid propellant strap-on boosters. *Project under development 1957–1963, then canceled.*

X-24A – Flew 28 times 1969–1971. Shape borrowed by X-38 CRV.

X-24B – Flying Flatiron Modified from X-24A. Double-delta planiform. *Flew 1973–1975.*

X-1 – Bell research rocket aircraft, first manned vehicle to break the sound barrier, October 14, 1947.

X-2 – Bell rocket-research aircraft, the first to reach Mach 3.

X-15 – Joint NASA/Department of Defense rocket-powered research vehicle; reached 67 miles altitude and Mach 6.7 in the early 1960s, and provided data for Mercury, Gemini, Apollo, and Space Shuttle programs. *Flew 199× 1959–1968.*

X-30 NASP – The National Aerospace Plane was to have used scramjet engines in a lifting body waverider to achieve single-stage-to-orbit spaceflight and rapid flight between major cities. *Conceptual early 1990s.*

X-33 – Subscale demonstrator for the VentureStar SSTO spaceplane. X-33 was an unmanned VTHL lifting body with integral composite propellant tanks and linear aerospike engines. The project was canceled in 2001 because of tank construction problems and concerns of unmanned test flights over the continental United States.

X-43 Hyper X – Unmanned scramjet vehicle, launched by B-52 mothership and Pegasus booster; demonstrated hypersonic cruise at speeds between Mach 6 and 10. *Flew 2× 2004.*

Index

Printed in the United States